Computational Analysis of Microscopy Images

Elvira Gagniuc

Computational Analysis of Microscopy Images

A Veterinary Medicine Perspective

Springer

Elvira Gagniuc
Department of Anatomical Pathology
University of Agricultural Sciences and Veterinary Medicine
Bucharest, Romania

ISBN 978-3-031-86269-4 ISBN 978-3-031-86270-0 (eBook)
https://doi.org/10.1007/978-3-031-86270-0

This Springer imprint is published by the registered company Springer Nature Switzerland AG
The registered company address is: Gewerbestrasse 11, 6330 Cham, Switzerland

I dedicate this book to my family and my mentors.
With profound gratitude,

Elvira

Foreword

The integration of computational analysis and machine learning with veterinary medicine has the potential to revolutionize diagnostic and prognostic approaches in veterinary pathology. This comprehensive book, *Computational Analysis of Microscopy Images: A Veterinary Medicine Perspective*, provides a detailed exploration of this transformative potential by combining traditional microscopy techniques with advanced computational methods to improve diagnostic accuracy, speed, and reproducibility. Veterinary medicine, like all branches of medicine, relies heavily on accurate diagnostic methods to make informed decisions about patient care. Segmentation and feature extraction, key areas discussed in the book, are fundamental in identifying and quantifying various features of interest in histological images. Through advanced thresholding techniques, region-based segmentation, morphological operations, and blob detection, this book demonstrates how specific features such as cell nuclei, abnormal structures, and other histopathological features can be accurately segmented and analyzed. The importance of quantitative image analysis addresses methods for counting cells and structures, for measuring morphological parameters, converting pixel data to micrometer scales, and quantifying staining intensities. These quantitative approaches are essential for providing an objective basis for diagnosing conditions such as malignancies, where abnormalities in cell size, shape, and distribution can serve as critical diagnostic markers. The advent of machine learning and artificial intelligence (AI) has brought about a paradigm shift in veterinary pathology. The book explores various machine learning techniques, focusing on convolutional neural networks (CNNs), transfer learning, and deep learning architectures specifically designed for veterinary pathology applications. These AI models can classify histological images, detect tumors, and differentiate between normal and abnormal tissues with high accuracy, as shown in several case studies presented in the book. The use of explainable AI further enhances model interpretability, making these advanced methods more accessible and practical for veterinary pathologists. However, the book does not shy away from discussing the challenges and limitations that accompany the integration of computational analysis and AI into veterinary pathology. Issues such as standardization of imaging data, anatomical and physiological variability, and balancing cost and benefit are thoroughly examined. This work serves as both a comprehensive guide and a call to action for the veterinary community to embrace computational analysis and AI technologies. As we stand at the frontier of a new era in veterinary diagnostics, the knowledge contained within these chapters equips veterinary pathologists, researchers, and clinicians with the tools needed to harness the power of computational methods to enhance the precision, efficiency, and reliability of veterinary diagnostics. The work compiled in this book is a testament to the collaborative efforts of pathologists, data scientists, and veterinarians dedicated to advancing veterinary medicine through innovation and technology. It is my hope that this book will inspire further research, development, and application of computational tools in veterinary pathology, ultimately improving animal health and welfare worldwide.

Manuella Militaru
Head of Department: Pathological Anatomy
University of Agronomical Sciences and Veterinary
Medicine of Bucharest, Faculty of Veterinary Medicine
Corresponding member of Academy of Agricultural
and Forestry Sciences "Gheorghe Ionescu Sisesti"
Bucharest, ROMANIA

Acknowledgments

I would like to thank Dr. Paul A. Gagniuc for the technical support and the *ELTA 90 Medical Research* company through the representative of Mr. Adrian Dragne for the access to the Nikon AX R confocal microscope with NSPARC technology (the last pictures in the appendices). I also want to thank Prof. Dr. Gherghiceanu Mihaela, for the access to the electron microscope at the *National Institute of Pathology Victor Babeş*.

Contents

About the Author

Elvira Gagniuc graduated from the *Faculty of Veterinary Medicine* (FMV) in Bucharest in 2010, completed her master degree in Veterinary Medicine with a specialization in *Laboratory Diagnostics* in 2012, and earned her Ph.D. in Veterinary Medicine in 2014. In 2013, as part of her doctoral scholarship, she benefited from advanced studies in the field of veterinary pathology within the Department of Pathology at the Faculty of Veterinary Medicine in Liège, Belgium. There, she acquired new diagnostic techniques in veterinary pathology (immunohistochemical diagnostics, PCR, electron microscopy). Currently, she is a lecturer in the *Department of Pathological Anatomy* at FMV Bucharest, teaching the general and special Pathology course and laboratory in the *French* section. In 2012, she joined the *Synevovet Laboratory* team, in the *Department of Pathological Anatomy*, and in 2019 became the coordinator of this department. She provides histopathological and cytopathological diagnostics for biopsies from both companion animals and farm animals. Dr. Gagniuc has participated in numerous specialization courses in *Belgium*, *Ireland*, and *Romania*; she annually attends various national and international symposia in the field of veterinary pathology and comparative medicine (in cities like Berlin, Paris, Lisbon, Athens, Cluj-Napoca, Bucharest, etc.) where she presents the results of her research studies. Elvira Gagniuc is a member of various national and international professional societies, including the *Davis-Thompson Foundation for the Advancement of Veterinary and Comparative Pathology*, the *European Society of Veterinary Pathology* (ESVP), and *Asociatia Generala a Medicilor Veterinari din Romania* (AGMVR). She has published over 75 scientific articles in various specialized scientific journals. Additionally, she has received six awards for the results of her published research, including four awarded by Unitatea Executiva pentru Finantarea Invatamantului Superior, a Cercetarii, Dezvoltarii si Inovarii (UEFISCDI). She is and has been a member of the research team for several scientific projects in the field of veterinary medicine. Areas of interest include histopathological, cytopathological, and immunohistochemical oncologic diagnostics in veterinary medicine, respiratory and digestive pathology in farm animals, toxicologic pathology, and fundamental science research in the field of comparative medicine.

Computational Analysis in Veterinary Medicine

1

Abstract

The integration of computational technologies into veterinary medicine is driving significant advancements in diagnostics, particularly in microscopy-based pathology. This chapter introduces the transformative role of computational analysis in veterinary medicine, emphasizing its ability to enhance the accuracy and efficiency of diagnostic processes. Microscopy, a fundamental tool in the diagnosis of animal diseases, traditionally relies on manual interpretation of complex tissue and cellular images. However, computational methods, such as machine learning and artificial intelligence, offer automated, precise image analysis that reduces human error and uncovers patterns not easily detected by the human eye. This evolution facilitates the detection of subtle pathological changes, improves reproducibility across laboratories, and accelerates diagnostic timelines. The chapter highlights the importance of computational analysis in revolutionizing veterinary diagnostics by providing a data-driven complement to clinical expertise. Additionally, it outlines the objectives and scope of the book, which explores advanced computational techniques and their applications in veterinary microscopy. This includes foundational knowledge of microscopy, digital imaging, and the application of machine learning, providing a comprehensive guide for veterinary professionals aiming to improve diagnostic precision and outcomes in their practice.

Keywords

Computational analysis · Veterinary medicine · Microscopy · Diagnostics · Image processing

E. Gagniuc, *Computational Analysis of Microscopy Images*,
https://doi.org/10.1007/978-3-031-86270-0_1

1.1 Introduction

In recent years, veterinary medicine has witnessed a transformative shift, driven by the integration of advanced computational technologies into traditional diagnostic practices. Among these innovations, computational analysis has emerged as a powerful tool that not only enhances the accuracy of diagnostics but also offers new insights into the complexities of animal health and disease. As veterinary professionals strive to provide better care for their patients, the role of computational methods in interpreting and analyzing vast amounts of data has become increasingly vital. This chapter serves as an introduction to the field of computational analysis within the context of veterinary medicine, with a particular emphasis on its application in microscopy—a cornerstone of diagnostic pathology. Microscopy, with its ability to reveal the intricate details of tissues and cells, is an indispensable tool in the diagnosis and study of lesions. However, the sheer volume and complexity of microscopic images often present challenges in manual interpretation. This is where computational analysis steps in, offering a means to automate, quantify, and enhance the evaluation of these images, thereby reducing human error and increasing diagnostic precision. The integration of computational analysis into veterinary medicine is not just a trend but a necessary evolution. With the advent of digital imaging technologies and the rapid growth of machine learning and artificial intelligence (AI), veterinary diagnostics are moving beyond the capabilities of the human eye. These technologies allow for the detection of subtle changes in tissue architecture, the quantification of cellular characteristics, and the identification of patterns that may be indicative of disease—all with unprecedented speed and accuracy. In this chapter, the significance of computational analysis in veterinary medicine is fully explored, tracing its development from early applications to the cutting-edge techniques used today. Also, the critical role of microscopy in veterinary diagnostics is discussed and how computational methods are revolutionizing this field. Furthermore, the objectives and scope of this book are outlined, providing a roadmap into various aspects of computational image analysis. The fusion of veterinary medicine with computational analysis marks a new era in diagnostics—one where data-driven insights complement clinical expertise, ultimately leading to better outcomes for animals and the people who care for them.

1.2 Overview of Veterinary Medicine

Veterinary medicine is a branch of science that deals with the prevention, diagnosis, and treatment of disease, disorder, and injury in animals. Covering a wide range of animal species, both domesticated and wild, the field encompasses various types of care and specialty areas including preventive healthcare, pharmacology, surgery, and internal medicine [1]. The practice not only contributes to animal health and welfare but also has significant implications for human public health by controlling zoonotic diseases (those that can be transmitted from animals to humans) and ensuring the safety of food supplies [2]. The discipline is rooted in both ancient and modern scientific practices, adapting over centuries as different societies recognized the need to care for animals that provided labor, companionship, and food [3]. Today, veterinary medicine is supported by advanced technologies such as digital imaging and complex biochemical analysis, similar to those used in human medical practices [4]. This evolution has led to specialized fields within veterinary medicine such as oncology, neurology, and cardiology, indicating the depth and breadth of the field. In contemporary practice, veterinarians work in private clinics, hospitals, regulatory bodies, and even in research and development, contributing to advances in medical technology, surgical techniques, and pharmaceuticals that benefit both animal and human health [5]. The profession demands a deep understanding of biological and veterinary sciences, a high level of skill in surgical procedures, and a compassionate approach to animal care and client communication. As veterinary medicine continues to evolve, it increasingly emphasizes holistic and preventive care approaches, underscoring the interdependence of animal and human health and the environment [6]. This holistic view forms a cornerstone of modern veterinary practice, reflecting a shift toward sustainable, globally aware strategies for medical care [6].

1.3 The Role of Computational Analysis

Computational analysis plays a critical role in veterinary medicine, particularly in enhancing the diagnostic capabilities through advanced imaging and data analysis techniques [7]. Veterinarians and researchers can achieve more accurate, detailed, and faster diagnoses, by using an integration of computational methods with traditional microscopy, which are crucial for effective treatment planning. At its core, computational analysis involves the application of various algorithms and mathematical models to interpret and manipulate digital images obtained from microscopy [8]. This approach allows for the extraction of meaningful patterns and metrics that might not be readily apparent through manual examination alone. For instance, automated image segmentation can isolate specific structures such as cells and tissues, enabling precise measurements and morphological assessments that contribute to identifying pathological changes [9]. Furthermore, the use of machine learning and artificial intelligence in computational analysis has revolutionized the field by providing tools that can learn from data [10]. These tools are capable of recognizing complex patterns associated with various diseases, predicting outcomes based on historical data, and even suggesting diagnostic categories [11]. This capability is particularly beneficial in veterinary pathology, where subtle differences in tissue structure can indicate the early stages of a disease. Quantitative image analysis, another facet of computational analysis, quantifies features within an image, such as cell count, area measurements, and intensity of staining. This quantification is vital for research and diagnostics, offering a level of precision that subjective interpretation cannot match [12]. Moreover, computational analysis supports the development of more standardized diagnostic processes in veterinary medicine [13]. It enhances reproducibility and accuracy across different labs and examiners by a reduction in the variability associated with human interpretation. Lastly, the educational aspect of computational analysis cannot be overlooked [11]. It provides a dynamic platform for training veterinary professionals, offering them tools to visualize complex biological structures and understand disease mechanisms in a more interactive and detailed manner. This educational benefit extends to improving the skills of existing professionals, keeping them updated with the latest developments in diagnostic technologies.

1.4 Importance of Microscopy in Veterinary Diagnostics

Microscopy plays a crucial role in veterinary diagnostics by allowing veterinarians and researchers to visually investigate the microscopic structure of animal tissues and cells [14]. This fundamental tool enables the detection, identification, and analysis of various pathological conditions that are not visible to the naked eye. Through the detailed examination of cell morphology and tissue architecture, microscopy aids in the diagnosis of diseases, including infectious diseases, cancer, and metabolic disorders. In the field of veterinary medicine, different types of microscopy, such as light microscopy, fluorescence microscopy, and electron microscopy, offer unique insights that are essential for a comprehensive diagnostic approach [15]. For instance, light microscopy is routinely used to observe the general features of histological samples stained with dyes, helping in the identification of cellular patterns and abnormalities [16]. Fluorescence microscopy, on the other hand, can be used to tag specific components of a cell or tissue with fluorescent markers, providing detailed images of cellular processes and structures [17]. Electron microscopy offers even higher resolution, allowing veterinarians to view the ultrastructure of cells, which is vital for diagnosing viral infections and structural pathologies. Moreover, the advancement in digital imaging technologies and computational analysis has significantly enhanced the capabilities of traditional microscopy [18]. Digital microscopy, combined with image analysis software, allows for automated cell counting, detailed morphometric analysis, and complex quantitative assessments that are reproducible and scalable [19]. These technologies enable veterinarians to obtain precise diagnostic information quickly, leading to faster and more accurate treatment decisions. Nevertheless, the importance of microscopy in veterinary diagnostics lies in its ability to provide detailed and accurate visualizations of biological material at the cellular and molecular levels, facilitating early diagnosis and effective management of diseases in animals [20]. This capability is indispensable not only for routine checks but also for conducting advanced research in veterinary pathology that furthers our understanding of animal diseases.

1.5 Historical Perspective and Recent Advances

In the historical context of veterinary medicine, microscopy has long been a cornerstone of diagnostic processes [21]. From its rudimentary beginnings in the late sixteenth century, when the first compound microscopes were developed, to the sophisticated digital imaging technologies of today, the evolution of microscopy has been pivotal [22, 23]. Early microscopes offered limited magnification and resolution, making them somewhat impractical for detailed medical diagnosis [24]. However, by the nineteenth century, improvements in lens design and the introduction of staining techniques, such as those developed by Paul Ehrlich, significantly enhanced the ability to view and differentiate cellular structures, paving the way for their use in pathology [25–28]. In the twentieth century, the field of veterinary microscopy benefited immensely from the advent of electron microscopy, providing unprecedented resolution that allowed for the visualization of ultrastructural details [29, 30]. This period also saw the development of fluorescence microscopy, which became essential for studying the functions of various cellular components using specific dye markers. Recent advances in microscopy and computational analysis have revolutionized veterinary medicine further. Digital microscopy now allows for high-resolution images to be captured, stored, and analyzed with computer software, facilitating more precise and faster diagnoses [31]. Innovations such as confocal microscopy, which offers three-dimensional imaging capabilities, and live-cell imaging technologies have provided veterinarians and researchers with tools to observe biological processes in real time [32]. Moreover, the integration of artificial intelligence and machine learning with image processing has opened new avenues for diagnostic methodologies [7]. Algorithms can now recognize patterns in image data that might be too subtle for the human eye, predicting ailments from images with increasing accuracy [13, 33]. Recent developments in deep learning, particularly convolutional neural networks, have been leveraged to automate and improve the accuracy and speed of image analysis in veterinary pathology [18, 10, 34, 35]. These technological advancements not only enhance the accuracy of diagnoses but also significantly shorten the time between sample collection and diagnosis, leading to faster, more efficient treatment decisions [34]. As expected, these technologies continue to evolve, and they promise to bring even more profound changes to veterinary medicine, enabling practitioners to offer more precise, effective, and personalized treatments [9, 34].

1.6 Objectives and Scope

The objectives and scope of this book revolve around equipping readers with a comprehensive understanding of computational analysis in veterinary medicine, specifically through the lens of microscopy. The book aims to bridge the gap between traditional veterinary practices and modern computational techniques, offering a detailed exploration of how these technologies can enhance diagnostic capabilities. It targets both novices who are beginning their journey in computational veterinary pathology and seasoned professionals looking to deepen their knowledge of advanced image analysis techniques. Starting with the fundamentals of microscopy, the book discusses the various types of microscopy used in veterinary diagnostics, such as brightfield, fluorescence, and electron microscopy, and covers essential preparatory techniques like staining and sample preparation. This foundational knowledge sets the stage for more advanced topics, including digital image processing and the computational tools available to assist practitioners in analyzing microscopic images. A significant portion of the book is dedicated to computational image analysis, where readers will learn about algorithms and software tools that can be used to process, enhance, and interpret images. This includes traditional image processing techniques and newer, more sophisticated methods involving machine learning and artificial intelligence. These sections not only cover the theoretical aspects but also provide practical applications and case studies demonstrating how these techniques are applied in real-world veterinary practice. The book also explores the integration of these computational methods into daily veterinary practice, highlighting how they can lead to more accurate diagnoses and better patient outcomes. Advanced imaging techniques and their applications from nano- to macro scales are discussed to illustrate the vast potential of combining microscopy with computational analysis.

Fundamentals of Microscopy in Veterinary Pathology

2

Abstract

Microscopy remains an essential tool in veterinary pathology, offering critical insights into the cellular and subcellular structures of animal tissues. This chapter introduces the foundational principles of microscopy and highlights its importance in veterinary diagnostics, where early and precise identification of diseases is vital. Thus, the chapter provides a comprehensive understanding of how microscopic analysis enhances diagnostic accuracy. Various microscopy methods are discussed, including brightfield, phase-contrast, fluorescence, confocal, and electron microscopy, each with unique applications and advantages in veterinary medicine. Brightfield microscopy is fundamental for routine examinations, while advanced techniques like fluorescence and electron microscopy offer higher resolution and specificity for identifying diseases such as infections and cancer. The chapter also examines how microscopy aids in reproductive studies and veterinary research, contributing to better disease management and treatment outcomes. Additionally, challenges such as sample preparation quality, equipment costs, and the need for specialized training are discussed, alongside the integration of computational tools that enhance diagnostic capabilities. This chapter provides a roadmap for veterinary professionals to navigate the evolving field of microscopy and its expanding role in improving animal healthcare.

Keywords

Microscopy · Light microscopy · Staining techniques · Confocal microscopy · Histopathology

E. Gagniuc, *Computational Analysis of Microscopy Images*,
https://doi.org/10.1007/978-3-031-86270-0_2

2.1 Introduction

Microscopy has been a cornerstone of veterinary pathology for decades, providing unparalleled insights into the cellular and subcellular structures of tissues and organisms. In veterinary medicine, where early and accurate diagnosis is critical for the treatment and prevention of diseases in animals, microscopy serves as an indispensable tool. This chapter examines the fundamentals of microscopy, providing a comprehensive overview to the specific needs and challenges of veterinary pathology. Understanding the basics of microscopy is crucial for veterinarians and pathologists who rely on this technology to identify cellular abnormalities, diagnose diseases, and monitor the effects of treatment. The ability to observe tissues at high magnifications allows for the detection of subtle pathological changes that might not be visible with the naked eye, thus enabling more precise and informed clinical decisions. This chapter begins with an exploration of the fundamental principles of light microscopy, the most commonly used type of microscopy in veterinary pathology. It will guide you through the various techniques of sample preparation and staining, essential steps that enhance the contrast and visibility of specific cellular components. The various array of microscopy techniques available today are discussed, from traditional brightfield microscopy to advanced methods like fluorescence, confocal, and electron microscopy. Each technique has its unique advantages and applications, making it essential for professionals to understand when and how to employ them effectively. For instance, while brightfield microscopy is excellent for routine examinations, fluorescence microscopy is better suited for identifying specific proteins within cells, and electron microscopy provides the ultrahigh resolution needed to observe minute details at the molecular level. Additionally, the chapter addresses the specific applications of these microscopy techniques in veterinary pathology, highlighting how they are used in diagnosing a wide range of animal diseases, from infectious diseases to cancer.

2.2 Basics of Light Microscopy

Light microscopy is a fundamental technique used in both clinical and research settings to observe structures that are too small to be seen by the naked eye [36]. This technique utilizes visible light to illuminate the sample being examined and optical lenses to magnify the image [37]. The basic components of a light microscope include the eyepiece or ocular lens, one or more objective lenses, a stage to hold the sample, a light source, and a series of mirrors and condensers that focus the light through the sample [38]. The magnification of a light microscope is determined by the combined powers of the ocular and objective lenses [39]. For instance, an ocular lens that magnifies 10 times (10×) and an objective lens that magnifies 40 times (40×) together produce a total magnification of 400 times (400×) [40]. Microscopes may vary from simple designs with one lens to complex systems with multiple lenses designed to enhance resolution and contrast [41]. The resolution of light microscopy, or its ability to distinguish two close points as separate, is influenced by the quality of the lenses and the wavelength of the light used [42]. The typical resolution limit of standard light microscopy is about 200 nanometers, meaning it can differentiate two points if they are at least 200 nanometers apart [42]. One of the most significant advantages of light microscopy is its ability to view living cells and organisms, allowing researchers and clinicians to observe biological processes in real time [32]. However, its use is also common in viewing fixed and stained samples, which can provide more contrast and details about cellular structures and compositions [32, 42]. Different techniques within light microscopy, such as phase-contrast microscopy, darkfield microscopy, and fluorescence microscopy, exploit variations in light and sample properties to enhance image contrast and detail [39]. These methods have expanded the applications of light microscopy beyond simple visualization, making it an essential tool in disease diagnosis, scientific research, and even in educational settings [43].

2.3 Sample Preparation and Staining Techniques

In microscopy, sample preparation and staining techniques are critical for enhancing visibility and contrast of structures in biological specimens. This process begins with the collection of the sample, which might involve biopsy, swab, or excision [14]. Ensuring that the sample is representative of the tissue or organism under investigation is paramount. Once collected, the sample is often fixed to preserve its structure and prevent degradation [44]. Fixation, typically achieved using chemical agents like formaldehyde or glutaraldehyde, stabilizes proteins and cellular components [44, 45]. After fixation, the sample undergoes dehydration through a graded series of alcohol (ethanol) washes, which removes water from the tissue [46]. This step is crucial for preventing distortion during the embedding process, where the tissue is infiltrated with a medium such as paraffin wax or resin, facilitating thin sectioning [47]. Sectioning, performed with a microtome for paraffin-embedded tissues or a cryostat for frozen samples, allows for the cutting of thin tissue slices for mounting on microscope slides [48]. These slices are typically between 5 and 10 micrometers thick to allow light or electrons to pass through the sample effectively. Staining is the next crucial step, where dyes or antibodies are applied to the sections [49]. Stains bind selectively to cellular or tissue components, differentiating them based on chemical properties or antigenic specificity. Common stains include *hematoxylin and eosin* (H&E), which stains nuclei blue and cytoplasm pink, and *Gram stain* for differentiating bacterial species [48]. Immunohistochemistry (IHC) uses antibodies tagged with enzymes or fluorescent dyes to detect specific protein markers, crucial for diagnosing diseases and identifying cellular functions. Special staining techniques such as silver staining for reticular fibers, PAS stain for glycogen and mucosubstances in tissues, and *Masson trichrome* for connective tissue are also employed based on diagnostic needs [45]. For electron microscopy, heavy metals such as osmium tetroxide and uranyl acetate are used to provide contrast by scattering electrons [50]. In fluorescence microscopy, fluorescent dyes are used that emit light upon excitation [46]. This technique is particularly useful for visualizing molecular interactions and functions within cells [32]. Each of these staining methods allows researchers and clinicians to observe microscopic structures with greater detail and specificity, facilitating accurate diagnostics and research in veterinary medicine [39] (see appendix section for different types of microscopy images).

2.4 Overview of Different Microscopy Techniques

Microscopy has evolved into an indispensable tool in veterinary medicine, facilitating deep insights into cellular structures and tissue architectures critical for diagnosis and research [51]. The range of microscopy techniques available allows for the examination of biological specimens in various ways, each with its own set of advantages and applications [52]. Brightfield microscopy is the most traditional form and relies on light passing through the sample. While simple and effective, it requires that samples be stained to produce contrast, making it excellent for general staining procedures such as hematoxylin and eosin. Phase-contrast microscopy enhances contrast in transparent and colorless specimens by converting phase shifts in light passing through the specimen into brightness changes in the image. It is particularly useful for examining living cells and tissues where staining is not feasible. Fluorescence microscopy involves exciting fluorescent dyes within the specimen with a specific wavelength of light and then detecting the emitted light. This technique is highly sensitive and specific, allowing for the visualization of specific components of cells and tissues, such as proteins or organelles [53]. Confocal microscopy provides the capability to generate sharp images of thick specimens by using a pinhole to eliminate out-of-focus light. This technique is instrumental in creating high-resolution three-dimensional images of structures within cells and tissues. Electron microscopy, which includes both transmission electron microscopy (TEM) and scanning electron microscopy (SEM), uses beams of electrons instead of light to create images [50]. TEM provides detailed images of the internal structure of cells, while SEM gives detailed surface views. Both are crucial for observing fine cellular and subcellular structures beyond the resolution of light microscopes. These techniques collectively expand the capability of veterinary pathologists to diagnose diseases, understand pathological conditions, and conduct sophisticated biomedical research, each providing a unique window into the microscopic world of veterinary pathology (please see appendix section for images taken with different types of microscopy).

2.4.1 Brightfield Microscopy

Brightfield microscopy, despite its simplicity, presents opportunities for computational analysis mainly in enhancing contrast and image clarity through digital post-processing techniques [54]. It is one of the most straightforward forms of microscopy, where light is either transmitted directly through or reflected off the sample, making the specimen appear dark against a bright background. Computational algorithms can be applied to brightfield images to perform tasks such as automatic counting, morphometry, and classification of cells [55]. Techniques like image thresholding, edge detection, and color deconvolution are used to extract meaningful information from the data that are often not readily visible to the human eye [11].

2.4.2 Phase-Contrast Microscopy

Phase-contrast microscopy enhances the contrast in transparent and colorless specimens by converting phase shifts in light passing through the specimen into brightness changes in the image. This method is ideal for live-cell imaging without the need for staining [52]. Computational analysis plays a critical role in quantifying the phase shifts and converting them into measurable data. Advanced image reconstruction algorithms can further enhance these images, allowing for detailed analysis of cellular dynamics and morphology [9]. The computational approach helps in the automated detection and tracking of cells in time-lapse imaging, providing insights into biological processes without invasive staining [11].

2.4.3 Fluorescence Microscopy

Fluorescence microscopy involves exciting electrons in molecules to emit light of a longer wavelength [53]. This technique is vital for the specific and sensitive detection of cellular components and can be used to observe dynamic processes within live cells. Computational analysis is integral to fluorescence microscopy, particularly in the areas of image restoration, signal enhancement, and quantification of fluorescence signals. Algorithms for noise reduction, such as deconvolution and denoising techniques, are essential to improve image quality [35]. Additionally, computational methods are used to analyze time-lapse fluorescence data, facilitating the study of cellular processes in real time.

2.4.4 Confocal Microscopy

Confocal microscopy provides the ability to obtain serial optical sections from thick specimens using a pinhole to block out-of-focus light, which is pivotal for constructing three-dimensional reconstructions of the sample [52]. The role of computational analysis in confocal microscopy is significant, especially in image processing and 3D modeling. Algorithms for 3D reconstruction, volume rendering, and quantitative image analysis enable researchers to visualize and measure complex biological structures in three dimensions. Computational tools can automate the process of stitching together multiple z-stacks, allowing for detailed analysis of larger volumes and structures within biological tissues [51, 13].

2.4.5 Electron Microscopy

Electron microscopy achieves much higher resolutions and magnifications, revealing the ultrastructure of cells, organelles, and macromolecules [50]. Computational analysis in electron microscopy extends to image acquisition, where algorithms control electron beams and detectors to optimize image quality and resolution [50]. Advanced image processing techniques, including tomographic reconstruction, are used to assemble two-dimensional images into detailed 3D models. Machine learning algorithms are increasingly applied to classify structures within electron microscopy images, identifying patterns and anomalies at resolutions beyond human visual capabilities [10, 55]. Moreover, the above microscopy techniques, equipped with computational analysis, not only enhance image quality and resolution but also extend the boundaries of traditional microscopy by providing tools for automatic quantification, classification, and complex structural analysis [52]. This integration plays a critical role in advancing the field of veterinary medicine, offering deeper insights into cellular and molecular processes.

2.5 Microscopy Applications in Veterinary Pathology

Microscopy plays a pivotal role in veterinary pathology, offering a detailed view of the cellular structure and morphology of tissues, which is crucial for accurate diagnosis and research [56]. In veterinary medicine, microscopy is utilized across a variety of applications ranging from routine diagnostic procedures to complex research methodologies [57]. One of the primary uses of microscopy in this field is the diagnosis of lesions. Histopathological examination of tissues allows pathologists to observe the morphological changes induced by different pathologies [58]. This is essential for identifying infectious diseases, as many pathogens can be directly observed in tissue sections. Microscopy also aids in the diagnosis of parasitic infections, where the parasites or their eggs may be visible within the tissue. Cancer diagnostics is another critical area where microscopy is indispensable [59]. It helps in the classification of tumors based on their cellular characteristics and structure, guiding treatment decisions and prognostication. Additionally, microscopy is used to evaluate the margins of surgical resections to ensure complete removal of malignant tissues. Microscopy is also integral to veterinary reproductive studies. It assists in examining the structure and quality of sperm and oocytes, which is valuable for breeding programs, especially in the case of valuable livestock and endangered species. Moreover, microscopy contributes significantly to veterinary research, enabling scientists to explore cellular responses to diseases, treatments, and vaccines at a microscopic level [60]. This microscopic evaluation extends to the study of zoonotic diseases, where understanding the progression of the disease in animals can provide insights into human infections.

In educational settings, veterinary students and trainees use microscopy to learn about normal and pathological tissue structures, building a foundational understanding that is crucial for their future clinical and diagnostic work [61]. Also, microscopy is being integrated with digital technology to enhance its applications in veterinary pathology. Digital microscopy allows for easier sharing of images for consultation and education, and when combined with image analysis software, it provides tools for quantifying and automating certain diagnostic tasks [62].

2.6 Challenges and Limitations in Veterinary Microscopy

In exploring the challenges and limitations in veterinary microscopy, several factors stand out that can impact the effectiveness and reliability of diagnostic results. One primary challenge is the quality of the sample preparation, which is critical for achieving clear and accurate images [63]. Poorly prepared samples can lead to artifacts that may obscure or mimic pathological features, complicating diagnosis [64, 65]. Another significant limitation is the resolution of standard optical microscopes, which may not be sufficient to detect subtle changes at the cellular or molecular level. This is especially pertinent in veterinary medicine, where different species present unique histological features that may require higher magnification and resolution for accurate identification. The cost and accessibility of advanced microscopy equipment also pose substantial barriers, particularly in less resourced settings. High-end tools like electron microscopes and advanced imaging software are expensive and require specialized training, which may not be feasible for all veterinary practices. Additionally, there is the issue of observer variability, which can lead to inconsistencies in how different technicians or pathologists interpret the same microscopic images [66]. This variability can be compounded by the subjective nature of many diagnostic processes in histopathology, where subtle differences in tissue appearance may lead to different diagnostic outcomes. Furthermore, the integration of computational analysis tools with traditional microscopy practices presents its own set of challenges, including the need for robust algorithms that can handle the variability and complexity of veterinary samples [67]. Computational tools must be carefully calibrated and validated to ensure they provide accurate and consistent results, which requires ongoing maintenance and updates as new diagnostic technologies and methods develop. Lastly, maintaining the pace of technological advancements in microscopy within the field of veterinary medicine requires continuous education and training [68]. Keeping up with new techniques, understanding the latest research, and implementing modern diagnostic tools demand significant time and effort from veterinary professionals [68].

3 Computational Image Analysis

Abstract

This chapter introduces the growing importance of computational image analysis in veterinary medicine, particularly in the analysis of microscopic images used in veterinary pathology. As diagnostic complexity increases, the integration of computational methods has become essential for veterinary practitioners. The chapter begins by outlining the basics of digital imaging and image acquisition, focusing on how images are captured, processed, and analyzed to extract valuable diagnostic information. Computational image analysis provides tools to handle large datasets, automating tasks such as cell counting, tissue segmentation, and feature extraction. These techniques enhance diagnostic accuracy and speed, enabling the detection of subtle patterns in histological images. The chapter also explores practical tools and software available for image analysis, including open-source platforms like ImageJ and CellProfiler, and advanced options like QuPath and HALO. Additionally, the chapter discusses the integration of computational methods into veterinary practice, emphasizing consistency in image acquisition and the role of machine learning. Case studies illustrate how these technologies improve diagnostic precision and workflow efficiency. The chapter aims to equip veterinary professionals with a solid foundation in computational image analysis, enabling them to apply these techniques to improve patient outcomes and streamline diagnostic processes.

Keywords

Digital imaging · Image processing · Machine learning · Segmentation · Feature extraction

E. Gagniuc, *Computational Analysis of Microscopy Images*,
https://doi.org/10.1007/978-3-031-86270-0_3

3.1 Introduction

In recent years, the field of veterinary medicine has increasingly used technology to enhance diagnostic accuracy, streamline workflows, and improve patient outcomes. Among the various technological advancements, computational image analysis has emerged as a critical tool, especially in the analysis of microscopic images used in veterinary pathology. As veterinary medicine continues to evolve, the integration of computational methods into traditional practices is not just an option but a necessity to keep pace with the growing complexity and volume of diagnostic data. This chapter provides an entry point into the world of computational image analysis, a discipline that lies at the intersection of computer science, mathematics, and veterinary pathology. It introduces the fundamental concepts of digital imaging, including how images are acquired, processed, and analyzed to extract meaningful information. Understanding these basics is crucial, as they form the foundation upon which more advanced techniques, such as machine learning and artificial intelligence, are built. Veterinary practitioners are increasingly required to handle large datasets of histological images, which demand precise and efficient analysis. Computational image analysis offers the tools and techniques necessary to process these images at scale, enabling the identification of subtle patterns that might be missed by the human eye. Computational methods not only enhance accuracy but also reduce the time required for diagnosis, by automating tasks like cell counting, tissue segmentation, and feature extraction. This chapter also explores the integration of computational analysis into veterinary practice, emphasizing the importance of these tools in daily diagnostic workflows. As this chapter progresses, insights will be gained into the various tools and software available for image analysis, along with practical examples of how computational techniques are transforming veterinary diagnostics. Therefore, the main purpose of this chapter is to provide a solid understanding of the basic principles of computational image analysis and how these principles can be applied to improve diagnostic accuracy and efficiency in veterinary medicine.

3.2 Basics of Digital Imaging and Image Acquisition

Digital imaging is the process of capturing and storing images in a digital format [69]. In the context of microscopy and veterinary medicine, digital imaging allows for the detailed examination, analysis, and storage of microscopic images, which can then be manipulated and analyzed using computational techniques [70]. The basic process involves capturing light from a specimen using an optical microscope and converting it into a digital image using a camera or a scanner [71]. The key components involved in digital imaging include the light source, the microscope, the camera or sensor, and the software used to process the images. The quality of the digital image depends on several factors, including the resolution of the camera, the quality of the optics, the lighting conditions, and the sensitivity of the sensor. High-resolution cameras are essential for capturing the fine details of microscopic specimens, which are often critical for accurate diagnosis and analysis in veterinary pathology. *Image acquisition* is the process of obtaining digital images from the microscope. This can be done using various methods, such as direct camera capture or scanning the specimen slide. Different types of cameras are used in digital microscopy, including charge-coupled device (CCD) cameras and complementary metal-oxide-semiconductor (CMOS) cameras. CCD cameras are known for their high-quality images and sensitivity, making them ideal for capturing detailed images in low-light conditions. CMOS cameras, on the other hand, are generally faster and more energy-efficient, though they may have lower sensitivity compared to CCD cameras. Proper image acquisition also involves optimizing the settings of the microscope and camera to achieve the best possible image quality. This includes adjusting the focus, exposure time, and white balance, as well as using appropriate filters and lenses. Image acquisition software plays a crucial role in controlling these parameters, capturing images, and managing the data. In veterinary medicine, digital imaging has revolutionized the way diagnostic information is gathered and analyzed. It allows for the precise measurement of cellular structures, the identification of pathological changes, and the sharing of images for remote consultations and collaborative diagnosis. Furthermore, digital images can be archived for future reference, enabling long-term studies and comparisons [72].

3.2.1 Consistency in Image Acquisition

Ensuring consistency in image acquisition is particularly crucial when utilizing machine learning for diagnostic purposes [73]. Machine learning models, especially those based on deep learning, are highly sensitive to the quality and characteristics of the input data. When images are acquired using different devices, there can be significant variations in resolution, color balance, noise levels, and overall image quality. These differences can affect the performance of the model, leading to inaccurate predictions or unreliable results [74]. Training a machine learning model on images from one acquisition device and then applying it to images from a different device can introduce discrepancies that the model was not trained to handle. For instance, a model trained on high-resolution images from a CCD camera may struggle to accurately interpret lower-resolution images from a CMOS camera. Even subtle differences, such as variations in lighting or slight changes in magnification, can impact the ability of the model to generalize from the training data to new, unseen images. This issue arises because machine learning models learn to identify patterns and features based on the specific characteristics of the training data. A lack of representativeness in the training data relative to real-world diagnostic cases can lead to a marked decline in the performance of the model [75]. This is known as the problem of domain shift, where the distribution of the training data differs from the distribution of the test data. To avoid this risk, it is essential to ensure that the images used for both training and diagnosis are captured under consistent conditions using the same acquisition device.

This consistency helps the model learn the relevant features without being confounded by variations in the imaging process itself. Additionally, maintaining uniformity in image acquisition across training and diagnostic stages can enhance the reproducibility of results, making the diagnostic process more reliable and robust in clinical practice [10]. Practitioners can maximize the effectiveness of machine learning models by carefully managing the image acquisition process, ensuring that they provide accurate and clinically relevant insights when applied to veterinary pathology [76]. In other words, when implementing machine learning models for diagnostic purposes in veterinary pathology, it is crucial to use images acquired from the same device for both training and diagnosis [77]. This consistency is essential because machine learning models, particularly those based on deep learning, learn specific patterns and features from the training data, which are influenced by the characteristics of the image acquisition device [78]. Variations in image quality, resolution, color balance, and noise levels between different devices can lead to a phenomenon known as domain shift, where the performance of the model significantly deteriorates when applied to images that differ from the training set [79]. For instance, a model trained on images captured with a high-resolution CCD camera might not perform well on images from a CMOS camera due to differences in image clarity and noise characteristics. These inconsistencies can result in reduced accuracy, unreliable predictions, and ultimately, incorrect diagnoses [80]. Maintaining uniformity in the image acquisition process ensures that the model generalizes well to new data, leading to more reliable diagnostic outcomes in clinical settings [81]. This need for consistency is well documented in the field of digital pathology, where studies have shown that even subtle differences in imaging conditions can affect the reproducibility and reliability of machine learning models used for diagnosis [82]. Therefore, for robust and accurate diagnostic applications, it is recommended to standardize the imaging process and use the same acquisition device for both training and application phases [83].

3.3 Key Concepts in Image Processing

Image processing is a critical field within computational analysis, particularly in the context of veterinary medicine. At its core, image processing involves the manipulation and analysis of digital images to enhance their quality, extract meaningful information, and enable accurate interpretation. Several fundamental concepts underpin image processing. One of the most basic concepts is *pixel representation*. Images are composed of pixels, which are the smallest units of a digital image. Each pixel has a specific intensity or color value that contributes to the overall image. Understanding how pixels form an image is crucial for further processing tasks. *Spatial resolution*, on the other hand, refers to the amount of detail in an image and is determined by the number of pixels. Higher resolution images contain more detail but require more storage and computational power. The bit depth of an image indicates the number of possible intensity levels per pixel, with higher bit depths allowing for more nuanced color or grayscale images. *Histogram analysis* is another key concept, providing a graphical representation of the distribution of pixel intensities in an image. One can enhance image contrast and adjust brightness levels by analyzing histograms and making features more discernible. Image filtering techniques, such as *convolution operations*, are used to enhance or suppress certain features within an image. Filters can smooth images (reducing noise) or highlight edges and boundaries, which are critical in identifying structures in veterinary pathology slides. Also, *segmentation* is a process where an image is divided into regions or objects to simplify its analysis. Techniques like thresholding, region growing, and edge detection are commonly used for segmentation in image processing. Segmentation is particularly important in identifying cells, tissues, or other structures within an image. Morphological operations such as dilation, erosion, opening, and closing modify the structure of objects within an image. These operations are typically used after segmentation to refine the shapes of detected structures, aiding in more accurate analysis. Feature extraction involves identifying and quantifying specific properties or characteristics of objects within an image. Features might include texture, shape, size, and color, which are critical for tasks such as classification or pattern recognition. Finally, *image transformation techniques*, such as *Fourier transforms*, allow the conversion of an image from one domain to another (e.g., from spatial to frequency domain), enabling further analysis and filtering that might not be possible in the original domain. Understanding these key concepts is essential for effectively analyzing and processing images in veterinary medicine, where accurate interpretation of microscopy images is vital for diagnosis and research, as the examples will show in the next chapters.

3.3.1 Image Processing vs Veterinary Pathology

Ordinary images and microscope images have different properties. Therefore, the question is: How can image processing be used and adapted for microscope images, especially for different tissues found in multiple species? Well, in the field of veterinary pathology, where precise and detailed image analysis is crucial, the application of advanced image processing techniques is indispensable [84]. These techniques enable the extraction of critical information from microscopy images, facilitating accurate diagnoses and in-depth research [85]. The following sections probe into key concepts in image processing, highlighting their importance and providing practical examples of how they enhance the analysis of histological slides in veterinary medicine. Pathologists can significantly improve their ability to interpret complex tissue structures and cellular patterns by understanding and applying these concepts [86, 87]. But let us see what these simple concepts are in more detail.

Pixel representation and understanding the distribution of pixel values are foundational when dealing with histological slides. For instance, in hematoxylin and eosin (H&E)-stained slides, each pixel color corresponds to different tissue components—hematoxylin stains cell nuclei blue, while eosin stains the extracellular matrix and cytoplasm pink. Accurately capturing and interpreting these pixel values is essential when distinguishing between normal and abnormal tissue structures.

Spatial resolution directly impacts the clarity with which microscopic structures are viewed. For example, when analyzing a sample for the presence of cellular infiltrates in a case of inflammation, a higher spatial resolution will allow the pathologist to distinguish individual cells and their morphology, such as the presence of lymphocytes or neutrophils. However, in routine pathology, a balance must be struck between resolution and processing efficiency, as extremely high-resolution images can be computationally intensive to analyze.

In veterinary pathology, **bit depth** becomes crucial when dealing with subtle variations in tissue staining. For example, in immunohistochemistry (IHC), which is used to detect specific antigens in tissues, higher bit depths can capture finer variations in staining intensity, leading to more accurate quantification of antigen expression levels. This is particularly important when comparing the expression of a marker between normal and cancerous tissues, where slight differences could be diagnostically significant [88].

Histogram analysis proves beneficial in cases where image contrast needs enhancement, such as in poorly stained or overexposed slides. For instance, when viewing a liver biopsy stained for glycogen using the periodic acid-Schiff (PAS) method, adjusting the histogram can help enhance the visibility of glycogen deposits within hepatocytes, making the analysis more reliable. Histogram equalization can bring out these subtle differences, aiding in more accurate diagnoses.

Image filtering is essential when preparing images for further analysis or interpretation. In veterinary pathology, filters like the *Gaussian blur* can be used to reduce noise in blood smear images, which is crucial when counting red and white blood cells. On the other hand, edge detection filters such as the *Sobel* or *Canny* methods can be applied to highlight the contours of structures like the glomeruli in kidney biopsies, allowing for the assessment of their integrity in cases of suspected glomerulonephritis.

Segmentation is particularly useful in isolating specific structures within a microscopy image. For instance, in veterinary oncology, segmenting tumor regions from surrounding healthy tissue is critical for assessing tumor margins and invasiveness [89]. Techniques like thresholding can help differentiate between necrotic and viable tumor areas based on differences in staining intensity. Similarly, region growing algorithms can be used to identify and segment regions of fatty infiltration in muscle tissue samples, which is important in diagnosing conditions like muscular dystrophy.

Morphological operations refine segmented structures, making them more suitable for further quantitative analysis. In the case of identifying parasitic eggs in stool samples, erosion and dilation operations can help isolate the eggs from background debris. After initial segmentation, applying opening and closing operations can smooth the boundaries of detected eggs, making them easier to count and classify accurately [90].

Feature extraction is critical when distinguishing between similar-looking structures in veterinary pathology. For example, when analyzing skin biopsies for different types of skin tumors, *extracting shape features can help differentiate between benign and* *malignant growths*. **Texture analysis**, using techniques like *gray-level co-occurrence matrix* (GLCM), can be particularly useful in differentiating between fibrous and glandular tissues, which may appear similar under certain staining conditions but have different underlying textures.

Finally, **image transformation techniques**, such as *Fourier transforms*, can be applied to detect *periodic patterns* in tissue structures that might not be obvious in the spatial domain. This is useful, for instance, in detecting the striations in muscle fibers or the regular arrangements of cells in certain types of tumors, such as fibrosarcomas. By transforming the image into the frequency domain, these patterns can be amplified, making them easier to analyze and quantify.

These image processing techniques, when applied to veterinary pathology, may greatly enhance the ability to detect, quantify, and interpret subtle changes in tissue and cell morphology, leading to more accurate diagnostics and a deeper understanding of disease processes. Each technique has specific applications that, when combined, provide a comprehensive toolkit for the analysis of complex histological images.

3.4 Tools and Software for Image Analysis

In this section, we will explore the various tools and software available for image analysis, particularly in the context of veterinary medicine. These tools range from open-source software to commercial packages, each offering unique features tailored to different aspects of image analysis. One of the most widely used tools is *ImageJ*, an open-source image processing program designed for scientific multidimensional images. It offers a broad range of plugins for tasks like *cell counting*, *image segmentation*, and *intensity measurement*, making it highly versatile for both novice and expert users. Another popular open-source software is *Fiji* (*Fiji* Is Just *ImageJ*), which is an enhanced version of *ImageJ*, bundled with many useful plugins preinstalled [91]. *Fiji* is especially favored for its ease of use and extensive documentation, which is beneficial for automating complex workflows.

CellProfiler is another powerful open-source tool specifically designed for analyzing images of biological specimens, such as cells [92]. It enables users to create pipelines that automate the extraction of quantitative data from images. This is particularly useful for high-throughput screening and can be integrated with machine learning tools to classify images based on extracted features.

For users requiring more advanced features, *QuPath* is an open-source software that offers powerful tools for digital pathology [93]. It supports *whole-slide image analysis*, allowing for the detection and quantification of tissue features across large images. The scripting capability of *QuPath* enables users to customize and automate complex analysis workflows, making it ideal for large-scale studies.

MATLAB is a commercial platform widely used for custom image processing applications [94]. With its Image Processing Toolbox, MATLAB offers extensive tools for developing algorithms that can be applied to a wide range of image analysis tasks. MATLAB is particularly useful for users who need to develop and test custom algorithms or who require advanced mathematical processing [95].

TensorFlow and *PyTorch* are two of the most prominent open-source deep learning frameworks used for developing machine learning models, including convolutional neural networks (CNNs) for image classification [96, 97]. Both frameworks offer robust libraries for image processing, feature extraction, and model training, making them ideal for integrating *AI* into image analysis workflows. *Keras API* in *TensorFlow* simplifies the development process, while *PyTorch* is known for its flexibility and ease of use in research.

HALO by *Indica Labs* is a commercial software that provides a comprehensive suite of tools for digital pathology, including tissue classification, multiplex IHC analysis (ImmunoHistoChemistry), and *AI*-driven image analysis [98]. *HALO* is designed for high-throughput and quantitative analysis of tissue images, making it a preferred choice for commercial and research laboratories.

In addition, *ZEN* by *Carl Zeiss* is a commercial software tailored for microscopy image acquisition and analysis [99]. It integrates seamlessly with *Zeiss* microscopes, providing tools for image stitching, 3D reconstruction, and advanced analysis of microscopy data. *ZEN* is widely used in both clinical and research settings for its high-quality image processing capabilities.

These tools and software provide a range of options for veterinary professionals and researchers, allowing them to perform detailed and accurate image analysis. Whether the need is for high-throughput screening, custom algorithm development, or integration of machine learning, the right tool or software can significantly enhance the analysis of histological images in veterinary medicine. Among the tools and software mentioned above, some are *libraries* used with programming languages, while others are *standalone* software applications designed for user interaction. In other words, *MATLAB*, *TensorFlow*, and *PyTorch* are *libraries* used with programming languages like Python, while *ImageJ*, *Fiji*, *CellProfiler*, *QuPath*, *HALO*, and *ZEN* are *standalone* software applications that offer a user interface for image analysis tasks.

Last but not least, *Olympus CellB*, with which all the images in this book were taken, deserves a special mention. Olympus CellB is a software platform designed for advanced imaging and analysis in various scientific fields, including histology, cell biology, and veterinary pathology. It offers a range of tools tailored for the acquisition, processing, and analysis of microscopic images, making it a valuable asset in the arsenal of tools used for image analysis in these domains. *CellB* is particularly noted for its user-friendly interface that combines powerful image acquisition capabilities with sophisticated analysis tools. It supports a wide range of Olympus microscopes and imaging devices, allowing seamless integration and control during image capture. This capability is crucial for ensuring that high-quality images are obtained, which can then be subjected to further analysis. One of the key features of *Olympus Cell^B* is its ability to perform various types of image processing, such as contrast enhancement, filtering, and noise reduction. These tools are essential for improving the clarity and quality of microscopic images, which can often be affected by factors such as uneven illumination or staining variability in histological samples.

Cell^B also includes advanced measurement and quantification tools that enable users to accurately assess different parameters within the images, such as cell size, shape, and density. These quantifications are crucial in research and diagnostic contexts where precise measurements are required to draw meaningful conclusions. In comparison with other *software like CellProfiler*, *HALO*, and *ZEISS ZEN*, *Olympus Cell^B* holds its own in terms of functionality and ease of use. *CellProfiler* is known for its open-source nature and flexibility in creating custom pipelines for complex image analysis tasks, making it highly suitable for high-throughput screening applications. *HALO*, on the other hand, is widely used in digital pathology, particularly for its strong emphasis on artificial intelligence and machine learning tools that automate the analysis of tissue samples, making it extremely valuable in veterinary pathology where large volumes of data need to be processed efficiently. *ZEISS ZEN* offers a comprehensive environment for both image acquisition and analysis, particularly excelling in 3D imaging and advanced microscopy techniques, and is favored in contexts where precision and integration with ZEISS hardware are essential. Therefore, *Olympus Cell^B* fits into this ecosystem of image analysis tools by providing a balanced mix of powerful features that are accessible to a wide range of users, from novice researchers to experienced professionals. Its integration with Olympus hardware ensures that users can take full advantage of their imaging systems, while its analysis capabilities are robust enough to handle the demands of scientific research and diagnostic workflows.

3.5 Integration of Computational Analysis in Veterinary Practice

The integration of computational analysis into veterinary practice represents a transformative approach to diagnostics and treatment. Veterinarians can enhance the accuracy and efficiency of their diagnostic procedures by simply using advanced image processing techniques and machine learning algorithms [75]. This chapter discusses the practical application of computational tools in everyday veterinary practice, thus exploring how these technologies can be seamlessly incorporated into existing workflows. Key topics include the use of automated image analysis software to assist in diagnosing diseases through the analysis of histological slides, the role of machine learning in predicting outcomes based on imaging data, and the integration of these tools with digital pathology systems [100]. This section also discusses the benefits of computational analysis, such as improved diagnostic precision, faster turnaround times, and the ability to handle large volumes of data, which is increasingly important in modern veterinary medicine. Practical case studies illustrate how veterinary practices have successfully implemented these technologies, highlighting the challenges and solutions encountered. Additionally, this chapter addresses the training and adaptation required for veterinary professionals to effectively use computational tools, ensuring that these technologies enhance rather than disrupt existing diagnostic and treatment processes. The ultimate goal is to provide a roadmap for integrating computational analysis into veterinary practices, leading to better patient outcomes and more efficient use of resources.

3.5.1 Anatomical and Physiological Variability

The integration of computational analysis in veterinary practice presents unique challenges and opportunities compared to its application in human clinical practice. While the fundamental principles of image processing and machine learning are similar across both fields, the multitude of cases in veterinary medicine, which encompasses a wide array of species, introduces complexities that are not typically encountered in human healthcare [101, 102]. One of the most significant differences lies in the anatomical and physiological variability between species [102]. In human medicine, computational models and diagnostic tools are generally optimized for a single species—humans—allowing for a high degree of standardization. In contrast, veterinary medicine must accommodate a vast range of species, each with distinct anatomical structures, tissue compositions, and disease manifestations. For instance, the computational analysis of a canine liver biopsy may require different parameters and training data than that of an equine or feline sample [103]. This variability implies the development of more flexible and adaptable computational models that can generalize across species or, alternatively, species-specific models that can be tailored to the particular needs of each animal type. Moreover, the variety of cases in veterinary practice also extends to the types of diseases encountered. Veterinary pathologists must diagnose conditions ranging from common infections and inflammations to rare genetic disorders, and even zoonotic diseases that can affect multiple species, including humans [100]. This variability means that computational tools must be equipped to handle a broader spectrum of conditions and be robust enough to recognize patterns that may only appear in certain species. For example, the presentation of lymphoma in a dog can be significantly different from that in a cat, not only in terms of cellular morphology but also in terms of disease progression and response to treatment [104]. Consequently, computational models must be trained on multiple datasets that reflect this range of pathology to ensure accurate diagnostics across species [105].

3.5.2 Standardization of Veterinary Imaging Data

Another critical consideration is the availability and standardization of veterinary imaging data. In human medicine, there are extensive, well-curated databases of medical images that can be used to train machine learning models [106]. These datasets are often standardized, with consistent imaging protocols and annotations, facilitating the development of reliable computational tools. In contrast, veterinary medicine faces challenges related to the heterogeneity of imaging data, which can vary widely depending on the species, the imaging equipment used, and the expertise of the operator. This variability can impact the quality and consistency of the data, making it more difficult to develop and validate computational models [105]. Addressing these issues may require collaboration across veterinary institutions to establish more standardized imaging protocols and create shared databases that represent a wide array of species and conditions [105].

3.5.3 The Balance Between Cost and Benefit

Additionally, the economic and practical realities of veterinary practice influence the adoption and implementation of computational analysis [107]. Veterinary clinics, especially smaller ones, may have limited access to advanced imaging technologies and computational resources compared to human hospitals. This disparity can affect the extent to which computational analysis is integrated into routine practice. For example, while a large veterinary teaching hospital might have the resources to employ AI-driven diagnostic tools and maintain extensive digital pathology archives, a small animal clinic may rely on more traditional methods [105]. Thus, computational tools for veterinary practice must be designed to be accessible and scalable, providing value even in settings with limited resources. Despite these challenges, the potential benefits of computational analysis in veterinary medicine are substantial. Professionals can achieve more accurate diagnostics, improve treatment planning, and ultimately enhance animal health outcomes, just by using specific computational tools to the specific needs of veterinary practice [102]. For instance, AI-powered tools that can quickly and accurately differentiate between benign and malignant masses in a wide variety of species could significantly improve the speed and accuracy of cancer diagnoses in animals, leading to earlier and more effective interventions [73, 79]. Nevertheless, while the core concepts of computational analysis are shared between veterinary and human medicine, the application in veterinary practice requires special consideration of the diversity of species, the variability of diseases, and the practical constraints of veterinary care settings [102, 105].

3.6 Python vs Veterinary Practice

The use of Python has become increasingly accessible, even for those who have little to no programming experience. This accessibility is largely due to the development of user-friendly libraries, tools, and environments that simplify the process of working with Python, making it an invaluable resource for veterinary practitioners. Whether a veterinarian is looking to analyze histological images, automate data processing, or integrate machine learning into their diagnostic workflow, Python offers a range of solutions that are both powerful and easy to use. One of the most significant advantages of Python is its similarity to MATLAB, a platform that many in the scientific and engineering communities are already familiar with. Like MATLAB, Python is designed to be intuitive and straightforward, allowing users to perform complex operations with just a few lines of code. This means that *veterinarians do not need to explore and learn the intricacies of programming to use Python effectively*. Instead, they can leverage the vast array of libraries available to accomplish tasks that would otherwise require extensive coding knowledge. The process of getting started with Python is relatively simple. Python is an open-source language, meaning it is freely available for download and use. In order to install *Python*, a practitioner can visit the official Python website and download the latest version suitable for their operating system. The installation process is straightforward, and during setup, there is an option to add Python to the system PATH environment variable, which makes it easier to run Python commands from any directory. Once Python is installed, the next step is to familiarize oneself with **PIP**, which stands for "Pip Installs Packages." PIP is a package manager for Python that allows users to easily install and manage the libraries they need. For instance, if a veterinarian wants to use a library for image processing, such as *OpenCV* or *scikit-image*, they can simply open a terminal or command prompt and type *pip install opencv-python* or *pip install scikit-image*. PIP handles the rest, downloading and installing the library along with any dependencies it requires. This process eliminates the need for users to manually track down and install each component, making it far more accessible for those with limited programming experience. The extensive library ecosystem of Python is one of the reasons it is so user-friendly. Libraries such as *NumPy*, *Pandas*, *Matplotlib*, and *scikit-learn* provide pre-built functions and tools that simplify complex tasks, from numerical analysis and data manipulation to machine learning and data visualization. For example, a veterinarian could use *Matplotlib* to create detailed plots of patient data with just a few lines of code or utilize *scikit-learn* to apply machine learning models to diagnostic tasks without needing to understand the underlying algorithms in depth. For those who prefer a more graphical interface or wish to avoid writing code altogether, there are also integrated development environments (IDEs like *Visual Studio Code*) and notebook interfaces like *Jupyter Notebook*. Jupyter allows users to write and execute Python code in a more interactive environment, where they can see the results of their code immediately. This environment is particularly beneficial for practitioners who may be experimenting with Python for the first time, as it provides a more engaging and less intimidating way to explore the capabilities of Python. Furthermore, there are countless tutorials, forums, and documentation available online that can guide practitioners through the process of using Python for various applications in veterinary medicine. Whether it is setting up a new Python environment, troubleshooting a piece of code, or learning how to apply a specific library, the community-driven nature of Python ensures that practitioners can find the support they need as they integrate this powerful tool into their practice.

4 Image Preprocessing Techniques

Abstract

Image preprocessing plays a pivotal role in the computational analysis of histological images, especially in veterinary pathology. This chapter introduces practical examples for various preprocessing techniques, which enhance image quality and ensure consistency for further analysis, such as segmentation, feature extraction, and classification. Key methods discussed include noise reduction, contrast enhancement, normalization, resizing, cropping, and data augmentation. Noise reduction methods like Gaussian blur and median filtering minimize artifacts without compromising diagnostic detail. Contrast enhancement, including histogram equalization, improves the visibility of subtle features in poorly contrasted images. Image normalization and calibration standardize images to account for variations in lighting, staining, and magnification, enabling accurate and reproducible measurements. Resizing and cropping optimize image handling and processing by focusing on regions of interest and ensuring consistent image dimensions. Data augmentation, particularly valuable in the context of limited veterinary datasets, simulates real-world variability through transformations like rotations and flips, improving model generalization and robustness. Together, these preprocessing techniques facilitate more precise and accurate computational analysis, ultimately contributing to better diagnostic outcomes in veterinary pathology.

Keywords

Noise reduction · Histogram equalization · Image normalization · Filtering · Contrast enhancement

E. Gagniuc, *Computational Analysis of Microscopy Images*,
https://doi.org/10.1007/978-3-031-86270-0_4

4.1 Introduction

Image preprocessing is a crucial step in the computational analysis of histological images, particularly in the context of veterinary medicine. It involves the application of various techniques to improve the quality and interpretability of images before they undergo more complex analysis. In microscopy-based diagnostics, where precision and accuracy are paramount, preprocessing ensures that the images are in an optimal state for subsequent tasks such as segmentation, feature extraction, and classification. Preprocessing serves several key functions. It can enhance the visual quality of images, making critical features more distinguishable. This is especially important in histology, where subtle variations in color, texture, and structure may be indicative of different pathological conditions. Additionally, preprocessing helps in standardizing images, correcting for artifacts and inconsistencies that may arise due to variations in sample preparation, staining, or microscope settings. There are some clear key objectives of image preprocessing, and those are related to *noise reduction, contrast enhancement, normalization and calibration, image resizing, and cropping* and *data augmentation.* Microscopic images often contain noise, which can obscure important details and lead to inaccurate analysis. Noise reduction techniques are applied to filter out unwanted artifacts while preserving the integrity of the true image content. Furthermore, histological images can sometimes suffer from poor contrast, making it difficult to differentiate between structures. Contrast enhancement methods, such as histogram equalization, are used to improve the visibility of key features, facilitating better analysis. Moreover, ariations in lighting conditions, staining intensity, and microscope settings can introduce biases in the image data. Normalization and calibration techniques are applied to standardize the images, ensuring that subsequent analysis is consistent and comparable across different samples. Also, to meet the input requirements of computational models or to focus on regions of interest, images may need to be resized or cropped. This step ensures that the images are in the appropriate format and size for further processing. In scenarios where the amount of labeled data is limited, data augmentation techniques can artificially expand the dataset by applying transformations such as rotations, flips, and shifts. This helps in improving the robustness and generalization capability of machine learning models. Properly preprocessed images lead to more accurate segmentation of tissues, better feature extraction, and more reliable classification outcomes. For example, reducing noise in a histological image might reveal the clear boundaries of a tumor or enhancing contrast could make it easier to count specific cell types. Moreover, the integration of computational tools in veterinary medicine demands a high level of consistency and reproducibility. Image preprocessing contributes to these goals by standardizing the inputs, making it possible to compare results across different studies, laboratories, or diagnostic setups. This is particularly important when images are analyzed by machine learning models, where variations in input data can lead to significant differences in performance.

4.2 Noise Reduction Methods

Noise reduction is a critical preprocessing step in image analysis, particularly in microscopy, where the quality of the images directly impacts the accuracy of subsequent analyses. In the context of histological images in veterinary pathology, noise reduction is particularly crucial due to the complexity and variability of tissue structures. Veterinary histology often involves examining a wide range of tissue types across different species, each with unique textural patterns and staining characteristics. The presence of noise can obscure critical diagnostic features such as cellular morphology, tissue boundaries, and staining intensity, leading to potential misinterpretation of the pathological state. For instance, when analyzing stained histological slides, noise can manifest as uneven staining or random speckles, which may interfere with the accurate identification of cellular structures. Various methods can be employed to reduce noise, like *Gaussian blur*, *median filtering*, *anisotropic diffusion*, *nonlocal means,* each with its own advantages and suitability for different types of images.

4.2.1 Gaussian Blur

Gaussian blur is one of the most common techniques for noise reduction [108]. It applies a *Gaussian* function to blur the image, effectively smoothing out high-frequency noise while preserving the overall structure of the image. This method is particularly useful when dealing with images that have a lot of fine, granular noise. *Gaussian blur*, while effective in smoothing these imperfections, must be applied carefully to avoid blurring out subtle but diagnostically significant details, such as the boundaries of tumor cells or the fine structure of nerve fibers.

Gaussian Blur

```
import cv2
from matplotlib import pyplot as plt

# Load the histological image!
# This image path:
#
# 'C:\\Users\\Elitebook\\Desktop\\his_img.jpg
#
# is used throughout
# this book for ease
# of handling the
# images:

image = cv2.imread('C:\\Users\\Elitebook\\Desktop\\his_img.jpg')

# Apply Gaussian Blur:

gaussian_blur = cv2.GaussianBlur(image, (31, 31), 0)

# Display the original
# and processed images
# side by side:

plt.figure(figsize=(10, 5))
plt.subplot(1, 2, 1)
plt.title('Original Image')
plt.imshow(cv2.cvtColor(image, cv2.COLOR_BGR2RGB))

plt.subplot(1, 2, 2)
plt.title('Gaussian Blur')
plt.imshow(cv2.cvtColor(gaussian_blur, cv2.COLOR_BGR2RGB))
plt.show()
```

Python implementation 1: Gaussian Blur.

The code from Python Implementation 4.1 is a practical example of how image processing techniques, like *Gaussian blur*, can be employed in veterinary histology to enhance the quality and interpretability of microscopic images, ultimately aiding in the accurate diagnosis and treatment of animal diseases. Histology plays a critical role in diagnosing various diseases, including cancers, infections, and other pathological conditions in veterinary patients. The ability to process and visualize these images effectively is crucial for accurate diagnosis and treatment planning. The code begins by importing essential libraries: *cv2* for image processing and *matplotlib.pyplot* for displaying images (Fig. 4.1). The histological image (non-suppurative encephalitis and lymphocytic-monocytic vasculitis) is loaded from the specified file path using *cv2.imread*, which reads the image in BGR format, as is standard with *OpenCV*. This image represents a stained brain tissue section from a pig that was diagnosed with encephalitis, non-suppurative, multifocal, moderate to severe, with perivascular lymphohistiocytic cuffing, neuronal degeneration, microgliosis, and perivascular hemorrhages [109, 110].

Next, the code applies a *Gaussian blur* to the loaded image using the *cv2.GaussianBlur* function. *Gaussian blur* is a technique used to reduce image noise and detail by averaging the pixels in a neighborhood defined by the kernel size, which is set to 31x31 in this case. In the context of veterinary histology, applying *Gaussian blur* can help in various scenarios, such as when there is a need to smooth the image for better visualization of overall

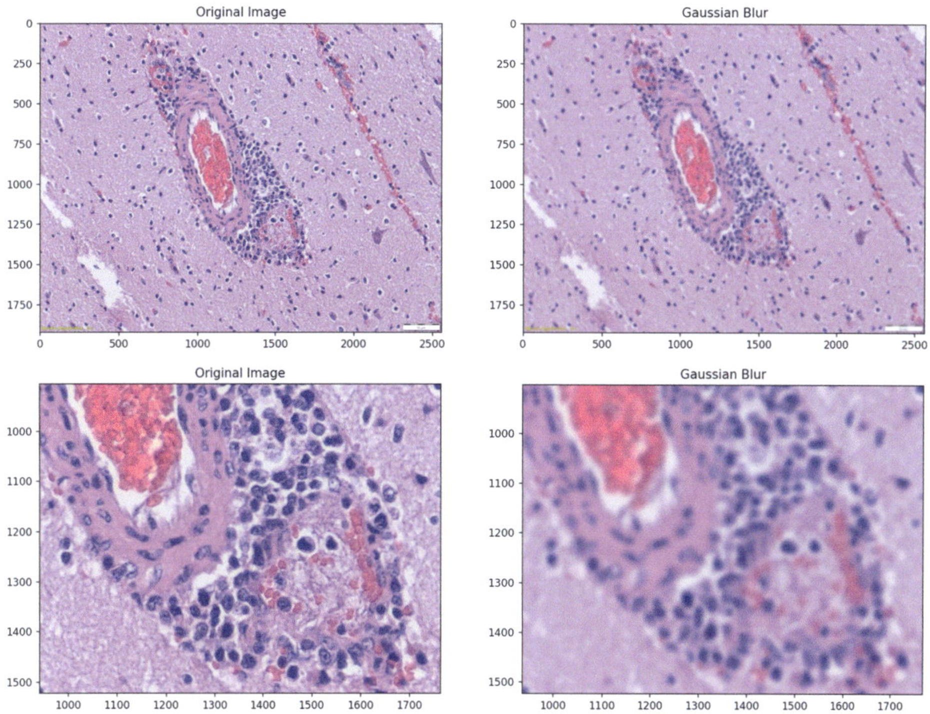

Fig. 4.1 Gaussian blur applied to an image representing brain tissue from a pig. The image shows encephalitis, non-suppurative, multifocal, moderate to severe, with perivascular lymphohistiocytic cuffing, neuronal degeneration, microgliosis, and perivascular hemorrhages. The image was captured at 20× magnification with a scale bar of 10 micrometers. The top-left image presents the original histological section, while the top-right image shows the same tissue after processing with Gaussian blur to enhance specific features. To better illustrate the effects of this processing, the bottom-left image provides a zoomed-in view of a region in the original image, and the bottom-right image presents a corresponding zoomed-in view of the same region in the processed image. This figure highlights the structural changes and features made more prominent through Gaussian blur processing, useful for emphasizing specific pathological details in histological analysis

tissue structures, or when preparing the image for further processing steps like *segmentation* or *edge detection*, which require noise reduction to function more effectively. The code further displays the original and blurred images side by side using *matplotlib.pyplot* (Fig. 4.1). The original image is shown in its natural form, preserving all details, while the blurred image provides a smoothed version that highlights broader tissue structures by reducing noise and minor details. In veterinary medicine, such visual comparisons can be crucial for pathologists who need to assess both fine and coarse details in tissue samples. For instance, the blurred image might be used to observe general tissue organization and structure, while the original image retains the details necessary for identifying specific cell types or pathological changes.

4.2.2 Median Filtering

In cases where histological slides exhibit "salt-and-pepper" noise, perhaps due to artifacts introduced during the staining process or digital acquisition, *median filtering* becomes particularly valuable [111]. This technique can effectively eliminate such noise while preserving the sharpness of cell nuclei and other critical structures, ensuring that these features are not mistaken for pathological alterations. Therefore, *median filtering* stands as another popular technique, especially effective for removing "salt-and-pepper" noise, which consists, as stated earlier, of random bright and dark pixels scattered across the image.

Median Filtering

```
import cv2
from matplotlib import pyplot as plt

# Load the histological image!
# This image path:
#
# 'C:\\Users\\Elitebook\\Desktop\\his_img.jpg
#
# is used throughout
# this book for ease
# of handling the images:

image = cv2.imread('C:\\Users\\Elitebook\\Desktop\\his_img.jpg')

# Apply median filtering:
median_filtered = cv2.medianBlur(image, 15)

# Display the original
# and processed images
# side by side:

plt.figure(figsize=(10, 5))
plt.subplot(1, 2, 1)
plt.title('Original Image')
plt.imshow(cv2.cvtColor(image, cv2.COLOR_BGR2RGB))

plt.subplot(1, 2, 2)
plt.title('Median Filtering')
plt.imshow(cv2.cvtColor(median_filtered, cv2.COLOR_BGR2RGB))
plt.show()
```

Python implementation 2: Median Filtering.

The median filter replaces each pixel value with the median value of the surrounding pixels, which helps in preserving edges while reducing noise. The Python code provided above is used for processing and analyzing histological images, which are essential in diagnosing and understanding various animal diseases. Python Implementation 4.2, begins by importing essential libraries, including *OpenCV* (*cv2*) for image processing and *Matplotlib* (*pyplot*) for visualizing the images (Fig. 4.2). The histological image, which could be a microscopic image of animal tissue, is loaded from a specified file path on the computer. Once the image is loaded, the code applies median filtering to the image. *Median filtering* is a technique used to reduce noise within an image, which is particularly important in histological analysis where the clarity of tissue structures is crucial for accurate diagnosis. Noise in an image can obscure important details, making it difficult to differentiate between normal and pathological tissues.

The median filter works by replacing each value of a pixel with the median value of the intensities in the surrounding neighborhood of that pixel. This process effectively reduces noise while preserving edges, which is beneficial in maintaining the integrity of the tissue structures in the image (Fig. 4.2) [109]. After applying the median filter, the original and filtered images are displayed side by side using *Matplotlib*. This visual comparison allows veterinarians or researchers to observe the effects of the noise reduction and ensure that the filtering process has not compromised important diagnostic features of the tissue.

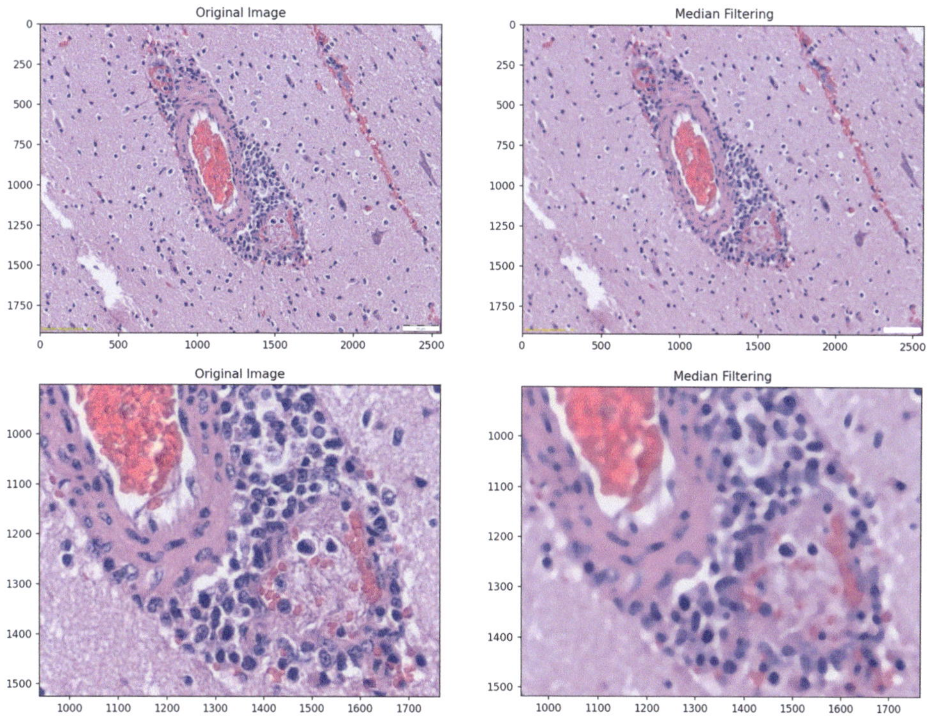

Fig. 4.2 Median filtering applied to an image representing brain tissue from a pig. The image shows encephalitis, non-suppurative, multifocal, moderate to severe, with perivascular lymphohistiocytic cuffing, neuronal degeneration, microgliosis, and perivascular hemorrhages. The image was taken at 20× magnification, with a scale bar representing 10 micrometers. The top-left image shows the original histological section, while the top-right image depicts the same tissue following median filtering, a processing technique used to reduce noise while preserving edges. For improved visualization of the processing effects, the bottom-left image offers a zoomed-in view of a specific region in the original image, whereas the bottom-right image provides a zoomed-in view of the same region in the processed image. The application of median filtering effectively highlights cell structures and vascular changes by smoothing out noise and enhancing important pathological details, aiding in the more accurate identification and interpretation of histological features

4.2.3 Anisotropic Diffusion

Anisotropic diffusion, also known as *Perona-Malik* filtering, is a more advanced noise reduction method that reduces noise while maintaining sharp edges [108]. This technique involves solving a partial differential equation that smooths the image in homogeneous regions but prevents diffusion across edges. It is particularly useful in medical imaging where edge preservation is crucial. Anisotropic diffusion is especially beneficial in veterinary histopathology where distinguishing between healthy and diseased tissue often relies on the integrity of tissue boundaries. For example, in the examination of a biopsy from a liver with suspected cirrhosis, this method can reduce background noise without compromising the visibility of fibrous septa, which are key diagnostic features.

Anisotropic Diffusion

```
import matplotlib
matplotlib.use('TkAgg')

# Above one may
# try 'Qt5Agg', 'Agg',
# 'MacOSX', etc.

from skimage import io, img_as_float
from skimage.restoration import denoise_tv_chambolle
from matplotlib import pyplot as plt

# Load the histological image!
# This image path:
# 'C:\\Users\\Elitebook\\Desktop\\his_img.jpg
# is used throughout
# this book for ease
# of handling the images:

image = img_as_float(io.imread(r'C:\Users\Elitebook\Desktop\his_img.jpg'))
print("Image loaded:", image.shape)

# Apply Anisotropic Diffusion
# without the 'multichannel'
# argument:

anisotropic_diffusion = denoise_tv_chambolle(image, weight=0.1)
print("Anisotropic diffusion applied")

# Display the original
# and processed images:

plt.figure(figsize=(10, 5))

plt.subplot(1, 2, 1)
plt.title('Original Image')
plt.imshow(image)
plt.subplot(1, 2, 2)
plt.title('Anisotropic Diffusion')

# cmap='gray' is needed
# if the image is grayscale:

plt.imshow(anisotropic_diffusion, cmap='gray')
plt.show(block=True)
```

Python implementation 3: Anisotropic Diffusion.

This script from Python Implementation 4.3 facilitates the enhancement of histological images, making it a valuable tool in the veterinary field for improving the accuracy and efficiency of diagnostic processes (Fig. 4.3). The code begins by configuring the *matplotlib* library to use the *TkAgg* backend, which ensures that the images will be displayed properly in a window on most operating systems. This is essential for visualizing the results of image processing in real time. The histological image, typically obtained from a tissue sample of an animal, is loaded into the program using the *io.imread* function from the *skimage* library. This image is then converted into a floating-point format with *img_as_float*, which is necessary for subsequent processing steps, as many image processing algorithms require the image data to be in a specific numerical format to perform calculations effectively.

The code then applies *anisotropic diffusion* to the image using the *denoise_tv_chambolle* function from the *skimage.restoration* module. *Anisotropic diffusion* is a sophisticated image processing technique that reduces noise while preserving important structural details, such as the edges of cells or tissue structures. This step is particularly valuable in veterinary histology, where the clarity of microscopic images is crucial for accurate diagnosis. Noise reduction helps pathologists focus on the relevant morphological features without the distraction of random variations that are not part of the actual tissue structure. Next, the code displays both the original and processed images side by side using *matplotlib* plotting capabilities. The original image is shown alongside the one that has undergone *anisotropic diffusion,* allowing for a direct comparison (Fig. 4.3). This visual comparison helps veterinarians and pathologists assess the effectiveness of the noise reduction process and ensures that the essential features of the tissue have been preserved. The code includes the option to apply a grayscale colormap (cmap='gray') to the processed image if it is grayscale, which is a common format for histological images.

4.2.4 Nonlocal Means

Nonlocal means (NLM) is another sophisticated noise reduction technique that works by averaging similar patches of pixels throughout the image, rather than just in the local neighborhood (Fig. 4.4). This method is highly effective in reducing noise while preserving fine details, making it ideal for high-resolution microscopy images [112]. NLM filtering is particularly suited for high-resolution histological images in which fine details are preserved, such as the texture of glandular structures in the gastrointestinal tract or the arrangement of muscle fibers in cardiac tissue. Moreover, NLM can suppress noise while maintaining the distinct textures that differentiate normal from abnormal tissue architecture by simply averaging similar patches across the image.

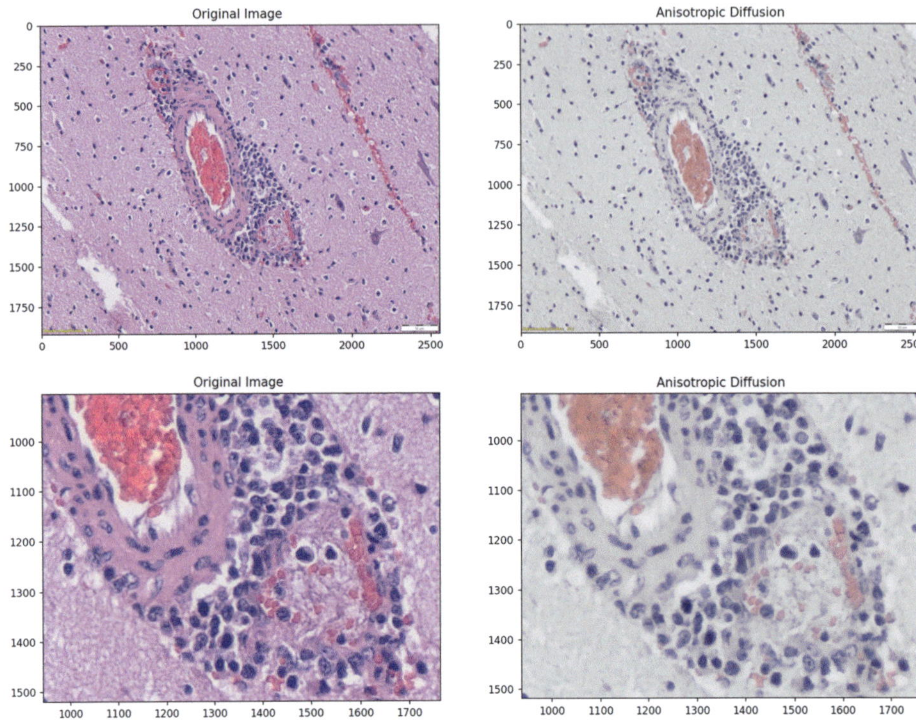

Fig. 4.3 Anisotropic diffusion applied to an image representing brain tissue from a pig. The image shows encephalitis, non-suppurative, multifocal, moderate to severe, with perivascular lymphohistiocytic cuffing, neuronal degeneration, microgliosis, and perivascular hemorrhages, which was captured at 20× magnification, with a scale bar indicating 10 micrometers. The top-left panel shows the original region from the histological section, whereas the top-right panel presents the same tissue after processing with anisotropic diffusion, a technique employed to enhance edges and smooth homogeneous areas by reducing noise while preserving significant structural information. To better visualize the effects of this processing technique, the bottom-left panel provides a zoomed-in view of a specific region from the original image, and the bottom-right panel shows a zoomed-in view of the same region in the processed image. The use of anisotropic diffusion effectively enhances the visibility of pathological changes, such as inflammatory infiltrates and vasculitic lesions, by improving the contrast between cellular structures and the surrounding tissue, facilitating more precise diagnostic interpretation

Non-Local Means

```
import cv2
from skimage import restoration, img_as_float
from skimage.transform import resize
from matplotlib import pyplot as plt
import time

################################
# Load the histological image:#
################################

image = cv2.imread(r'C:\Users\Elitebook\Desktop\his_img.jpg')
print("Image loaded:", image.shape)

# Convert image to RGB
# and float format:

image = cv2.cvtColor(image, cv2.COLOR_BGR2RGB)
image_float = img_as_float(image)
print("Image converted to float format.")

# Option 1: Further resize
# the image to speed up
# processing:

image_resized = resize(image_float, (image_float.shape[0] // 4, image_float.shape[1] // 4),
anti_aliasing=True)
print("Image resized to:", image_resized.shape)

# Option 2: Process a smaller region
# of interest (ROI) within the image!
# Uncomment the following lines to
# process only a smaller region
# ...
# image_roi = image_float[1000:1500, 1000:1500]
# print("Processing a smaller region of interest:", image_roi.shape)

# Apply Non-Local Means (NLM)
# Filtering on the resized image
# or replace 'image_resized' with
# 'image_roi' to use the smaller region:

start_time = time.time()

nlm_filtered = restoration.denoise_nl_means(image_resized, h=0.05,
                                             fast_mode=True, patch_size=3,
                                             patch_distance=1)

print("NLM filtering completed in: {:.2f} seconds".format(time.time() - start_time))

# Display the original
# and processed images
# side by side:

plt.figure(figsize=(10, 5))

plt.subplot(1, 2, 1)
plt.title('Original Image (Resized)')
plt.imshow(image_resized)

plt.subplot(1, 2, 2)
plt.title('Non-Local Means (NLM) Filtering')
plt.imshow(nlm_filtered)

plt.show()

print("Plot shown.")
```

Python implementation 4: Non-Local Means.

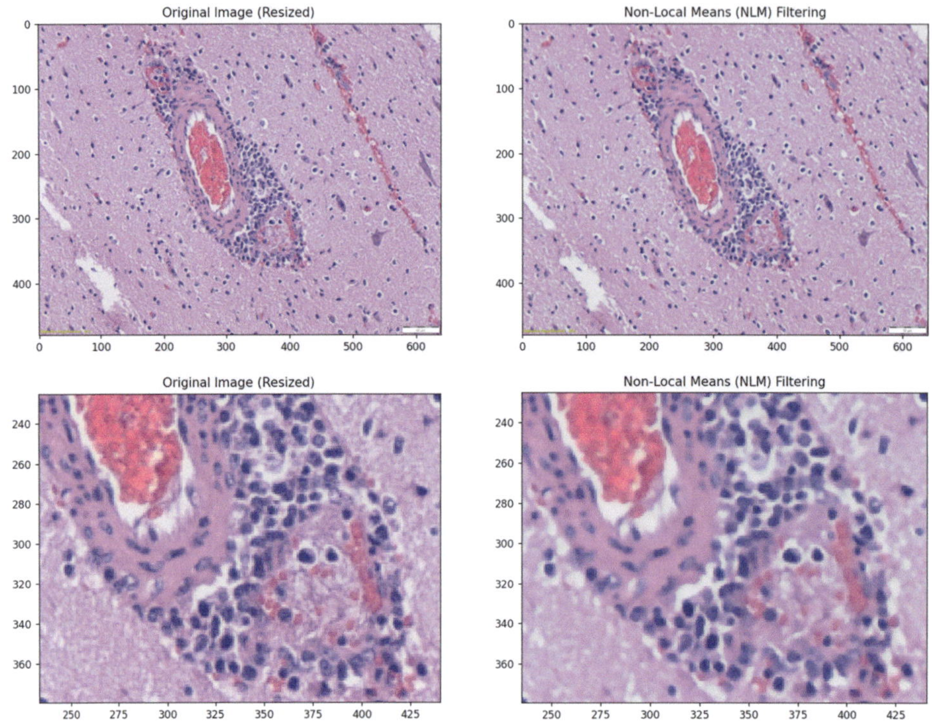

Fig. 4.4 Nonlocal means applied to an image representing brain tissue from a pig. The image shows encephalitis, non-suppurative, multifocal, moderate to severe, with perivascular lymphohistiocytic cuffing (Virchow-Robbins spaces contain numbers of lymphocytes and macrophages), neuronal degeneration, microgliosis, and perivascular hemorrhages. The image was taken at 20× magnification, with a scale bar representing 10 micrometers. The top-left panel displays the original, unprocessed image of the tissue. The top-right panel shows the same tissue after processing with the nonlocal means (NLM) algorithm, a technique designed to reduce noise while maintaining edges and important structural details. This processing enhances the contrast between different tissue components, helping to reveal subtle pathological features. To provide a clearer understanding of the impact of this processing technique, the bottom-left panel offers a zoomed-in view of a specific area in the original image, while the bottom-right panel presents a zoomed-in view of the same area in the processed image. The application of nonlocal means filtering effectively reduces background noise and enhances the clarity of inflammatory infiltrates and vascular changes, which are key features in diagnosing encephalitis and vasculitis, thereby aiding in more accurate histopathological evaluation

The code found in Python Implementation 4.4 is enhancing the quality of histological images used in diagnostic procedures. As already stated, histological slides, which are critical for diagnosing various conditions in animals, often contain noise and artifacts that can obscure important details. This noise can make it difficult for veterinary pathologists to accurately assess the cellular structures and tissue organization necessary for making a diagnosis. Therefore, processing these images to reduce noise while preserving crucial details is paramount. This code begins by loading a histological image of veterinary tissue using the *OpenCV* library. This image is initially in BGR format, which is standard for images handled by *OpenCV*. However, for more advanced image processing, the image is converted to the RGB format and then transformed into a floating-point representation. This conversion is necessary because the subsequent noise reduction algorithms work best with images in this format, allowing for more precise calculations during processing. Given that histological images are often high resolution, processing them in their original size can be time-consuming and computationally expensive. The code addresses this by resizing the image to a quarter of its original size (Fig. 4.4). This reduction in size decreases the number of pixels that need to be processed, thereby significantly speeding up the subsequent steps. However, it is important to note that while resizing speeds up processing, it also reduces the resolution, which might result in the loss of some finer details. This trade-off is often acceptable in veterinary diagnostics, where the primary goal is to enhance the clarity of major structures rather than focus on the minutiae. An alternative approach within the code allows for processing only a specific region of interest (ROI) within the image. This method is particularly useful when a veterinary pathologist is interested in examining a specific area of the tissue, such as a suspected lesion or a region showing abnormal cellular activity. The processing time can be further reduced while maintaining the original resolution of the area of interest by simply focusing on a smaller area (Fig. 4.4).

The core of the code is the application of the *nonlocal means* (NLM) filtering algorithm, another sophisticated method for noise reduction (Python Implementation 4.4). NLM filtering works by comparing each pixel to its surrounding pixels and averaging them based on their similarity, effectively smoothing out noise while preserving important details like edges and textures. This is crucial because it helps to clarify the tissue structure, making it easier for pathologists to identify abnormalities (Fig. 4.4). Again, it is important to note that NLM filtering is computationally intensive, particularly for high-resolution images. Even after resizing the image, the processing can still take a considerable amount of time, as demonstrated by the example, where it took over a minute to complete the filtering process on a resized image. This extended processing time is a common challenge when dealing with complex algorithms like NLM, especially when the quality of the output is prioritized over speed. After the filtering process is complete, the code displays the original resized image alongside the NLM-filtered image. This side-by-side comparison allows veterinary professionals to visually assess the effectiveness of the noise reduction and to determine whether the processed image provides a clearer view of the tissue structures necessary for diagnosis.

Image preprocessing is a foundational step in the computational analysis pipeline, ensuring that histological images are in their best possible form for further analysis. Each of these methods (i.e., *Gaussian blur*, *median filtering*, *anisotropic diffusion*, *nonlocal means*) has its place in the preprocessing pipeline, and the choice of method often depends on the specific characteristics of the noise present in the images and the goals of the analysis. In practice, it is common to experiment with different noise reduction techniques to find the most suitable one for a given set of images. However, applying noise reduction methods to histological images in veterinary pathology requires a nuanced approach, balancing the need to eliminate noise with the imperative to preserve diagnostically relevant details. Each method offers unique advantages that, when appropriately chosen, can enhance the clarity and accuracy of histological assessments, ultimately supporting more precise veterinary diagnoses.

4.3 Contrast Enhancement and Histogram Equalization

Contrast enhancement and histogram equalization are essential techniques in the field of veterinary histology for improving the visibility of features in histological images. These techniques are particularly valuable when dealing with images that suffer from poor contrast, which can obscure important details necessary for accurate diagnosis. In veterinary pathology, where the identification of subtle differences in tissue structure can be critical, enhancing contrast can make a significant difference in the clarity and interpretability of images. *Contrast enhancement involves adjusting the intensity levels of an image to increase the difference between the lightest and darkest areas.* Contrast enhancement, as a method, makes the features within the image more distinguishable by stretching or compressing the range of intensity values. For instance, in a histological slide stained to reveal cellular structures, contrast enhancement can help to make the boundaries of cells and tissues more pronounced, aiding in the identification of abnormal growths or lesions. This process is particularly useful in veterinary diagnostics, where the quality of the histological image can directly impact the accuracy of the diagnostic outcome. On the other hand, *histogram equalization* is a more advanced technique for contrast enhancement that works by redistributing the intensity values of the image to achieve a more uniform histogram. The histogram of an image represents the distribution of intensity values, and in many histological images, this distribution may be skewed, leading to a concentration of values in certain ranges, which results in poor contrast. Histogram equalization spreads out these intensity values more evenly across the entire range, thereby enhancing the overall contrast of the image. In the context of veterinary histology, this technique can be particularly beneficial for highlighting subtle variations in tissue density or staining intensity, which are often key indicators of pathological changes. When applied to veterinary histological images, these techniques can significantly improve the visibility of crucial details. For example, in images of stained tissue sections, contrast enhancement and histogram equalization can help to differentiate between various types of cells, highlight abnormal cell growth, and make the overall structure of the tissue more apparent. This improved visibility is not only crucial for manual examination by veterinary pathologists but also enhances the accuracy of automated image analysis algorithms that may be used to quantify certain features or identify patterns indicative of disease. However, it is important to consider the potential drawbacks of these techniques. Over-enhancing contrast or applying aggressive histogram equalization can introduce artifacts or exaggerate noise in the image, which could lead to misinterpretation. Therefore, in veterinary histology, these techniques must be applied with care, often requiring fine-tuning to balance between enhancing important details and maintaining the overall integrity of the image.

4.3.1 Contrast Enhancement

As stated above, contrast enhancement and histogram equalization are powerful tools in the enhancement of veterinary histological images, making critical details more visible and aiding in the diagnostic process. These techniques, when used judiciously, can greatly improve the quality of images, making them more informative and easier to interpret for both manual and automated analyses. The code from Python Implementation 4.5 is a practical example of how contrast enhancement can be applied to histological images, which are crucial in the field of veterinary medicine. This code leverages the *OpenCV* library to manipulate the contrast of an image, making it a valuable tool for improving the clarity and diagnostic utility of histological slides used by veterinary pathologists. The process begins with loading a histological image using the *imread* function from *OpenCV*. This image is initially in the BGR color format, which is standard in *OpenCV*. To prepare the image for further processing and visualization, it is converted to the RGB color format, which is more suitable for display and analysis in Python environments.

Contrast enhancement

```
import cv2
import numpy as np
from matplotlib import pyplot as plt

# Load the histological image:
image = cv2.imread(r'C:\Users\Elitebook\Desktop\his_img.jpg')
image_rgb = cv2.cvtColor(image, cv2.COLOR_BGR2RGB)

# Apply contrast enhancement:
alpha = 1.5  # Contrast control (1.0-3.0).
beta = 0     # Brightness control (0-100).

enhanced_image = cv2.convertScaleAbs(image_rgb, alpha=alpha, beta=beta)

# Display the original and
# contrast-enhanced images:

plt.figure(figsize=(10, 5))

plt.subplot(1, 2, 1)
plt.title('Original Image')
plt.imshow(image_rgb)

plt.subplot(1, 2, 2)
plt.title('Contrast Enhanced Image')
plt.imshow(enhanced_image)

plt.show()
```

Python implementation 5: Contrast enhancement.

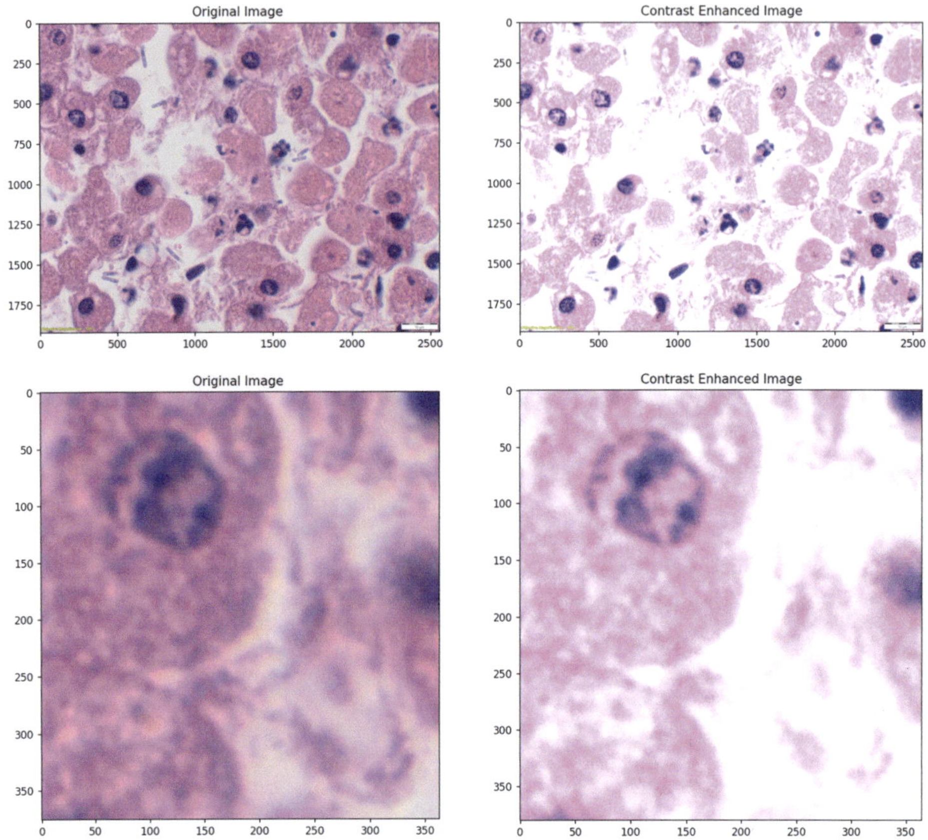

Fig. 4.5 Contrast enhancement for hepatic parenchyma of a pig infected with *Clostridium spp.* Images were captured at 100× magnification, with a scale bar representing 10 micrometers. The top-left panel shows the original histological image of the liver tissue, displaying the morphology of hepatocytes and areas of bacterial infection. The top-right panel presents the same image after contrast enhancement processing, which improves the visibility of structural details and highlights the infected areas. The bottom-left panel provides a zoomed-in view of the original image, focusing on the cellular and bacterial features. The bottom-right panel shows a zoomed-in view of the contrast-enhanced image, further emphasizing the changes in contrast to better visualize the infected regions and cellular structures

Once the image is in the correct format, the code applies contrast enhancement through a method that adjusts the intensity values of the pixels (Python Implementation 4.5 and Fig. 4.5). This is done using the *convertScaleAbs* function, where two key parameters are adjusted: *alpha*, which controls the contrast, and *beta*, which controls the brightness. In this example, *alpha* is set to 1.5, increasing the contrast of the image, while *beta* is left at 0, meaning that no additional brightness is added. The contrast enhancement process makes the dark regions of the image darker and the bright regions brighter, thereby increasing the overall contrast. This enhancement is particularly important in veterinary histology, where subtle differences in tissue structure and cellular organization need to be clearly visible to identify pathological changes (Fig. 4.5). From a veterinary perspective, the enhancement of contrast over the histological images can significantly aid in the diagnosis of diseases. Veterinary pathologists should rely on these images to identify abnormalities such as inflammation, neoplasia, or infectious agents in tissue samples. Thus, the code helps to make these features more distinguishable by enhancing contrast, especially in cases where the original image might be poorly stained or suffer from low contrast due to suboptimal sample preparation.

This can be particularly useful when dealing with complex or subtle histopathological changes that are not easily visible in the original image. The final step of the code displays the original and contrast-enhanced images side by side using *Matplotlib* (Fig. 4.5). This visual comparison may allow veterinary professionals to see the impact of the contrast enhancement and assess whether the enhanced image provides a clearer view of the tissue structures. In practice, such visual enhancements can improve the accuracy of diagnoses by making critical features more apparent, thereby supporting better clinical decision-making in veterinary medicine. This code (Python Implementation 4.5) not only demonstrates a technical approach to improving the quality of histological images but also highlights its importance in the veterinary field, where clear and detailed images are essential for accurate pathology assessment and diagnosis.

4.3.2 Histogram Equalization

This second code (Python Implementation 4.6) builds upon the principles of enhancing histological images for better visibility, much like the contrast enhancement code discussed earlier, but it introduces a more sophisticated technique: *histogram equalization*. Both codes share the common goal of improving the quality of histological images, which are critical in veterinary pathology for diagnosing diseases by examining tissue samples. However, they differ in the methods they use to achieve this improvement. In the first code seen in Python Implementation 4.5, contrast enhancement is achieved by directly manipulating the intensity values of the image using a simple linear transformation. This approach adjusts the contrast by scaling the intensity values, making darker areas darker and brighter areas brighter, which is particularly useful for quickly improving the visibility of features in a histological image that may appear washed out or lack contrast.

Histogram equalization

```
import cv2
from matplotlib import pyplot as plt

# Load the histological image:
image = cv2.imread(r'C:\Users\Elitebook\Desktop\his_img.jpg')
image_rgb = cv2.cvtColor(image, cv2.COLOR_BGR2RGB)

# Convert image to YUV color space:
image_yuv = cv2.cvtColor(image_rgb, cv2.COLOR_RGB2YUV)

# Apply Histogram Equalization
# on the Y channel (luminance):
image_yuv[:, :, 0] = cv2.equalizeHist(image_yuv[:, :, 0])

# Convert back to RGB format
hist_eq_image = cv2.cvtColor(image_yuv, cv2.COLOR_YUV2RGB)

# Display the original and
# histogram-equalized images:
plt.figure(figsize=(10, 5))

plt.subplot(1, 2, 1)
plt.title('Original Image')
plt.imshow(image_rgb)

plt.subplot(1, 2, 2)
plt.title('Histogram Equalized Image')
plt.imshow(hist_eq_image)

plt.show()
```

Python implementation 6: Histogram equalization.

The second code (Python Implementation 4.6), on the other hand, employs *histogram equalization*, a more advanced technique that improves contrast by redistributing the intensity values of the image. This method works by first converting the image from the RGB color space to the YUV color space. The YUV color space separates the image into its luminance (Y) and chrominance (U and V) components, allowing the histogram equalization to be applied specifically to the luminance channel. This targeted approach enhances the contrast of the image without affecting the color balance, which is crucial in maintaining the integrity of the histological stain colors that pathologists rely on for diagnosis (Fig. 4.6). After applying histogram equalization to the luminance channel, the image is converted back to the RGB color space for display. The resulting histogram-equalized image typically shows improved contrast across the entire range of intensities, making subtle details in the tissue structure more apparent.

This can be particularly beneficial in veterinary histology when dealing with images that have uneven illumination or when trying to bring out details in both the darker and lighter regions of the tissue. When comparing the two methods, histogram equalization can be seen as a more nuanced approach that not only enhances contrast but does so in a way that ensures a more uniform distribution of intensities across the image. This can lead to a more balanced and visually consistent image, making it easier to identify critical structures within the tissue sample. However, it is also more complex than the straightforward contrast adjustment method used in the first code. Therefore, in the context of veterinary medicine, both techniques are valuable. *Contrast enhancement* might be preferred

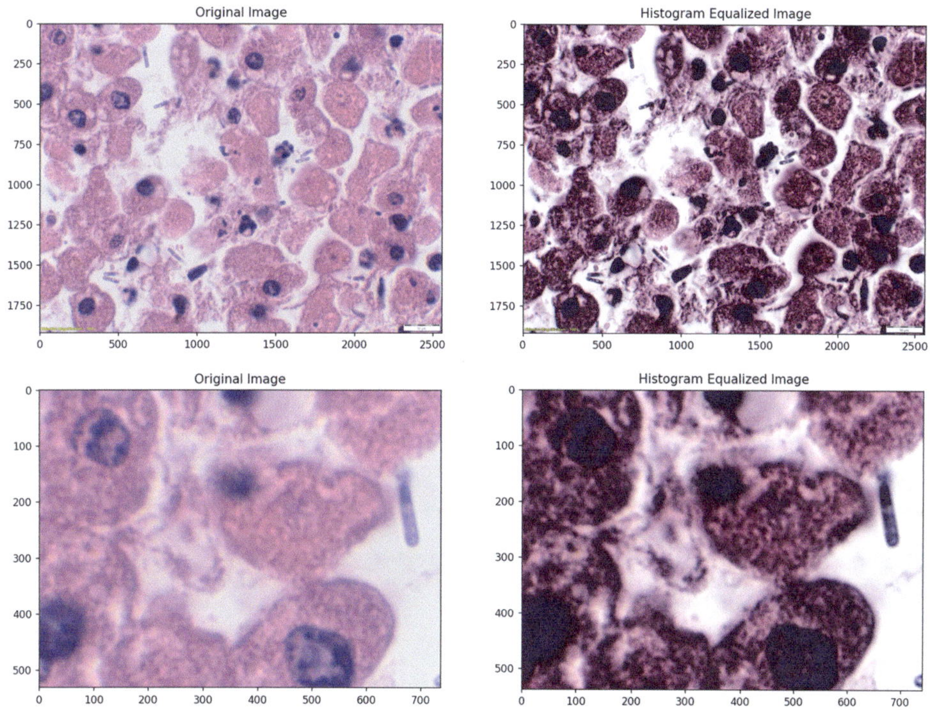

Fig. 4.6 Histogram equalization on liver tissue from a pig showing hepatic parenchyma infected with *Clostridium spp*. Images were captured at 100× magnification, with a scale bar representing 10 micrometers. The top-left panel displays the original histological image, revealing the presence of bacterial infection and liver cell structures. The top-right panel shows the image after histogram equalization processing, which enhances the contrast and detail of the histological features. The bottom-left panel provides a zoomed-in view of the original image, focusing on the affected hepatic parenchyma and bacterial colonies. The bottom-right panel presents a zoomed-in view of the processed image, demonstrating the improved visualization of the cellular and microbial structures achieved through histogram equalization

when quick adjustments are needed, or when dealing with images that only require a slight improvement in contrast. Histogram equalization, however, offers a more comprehensive enhancement, particularly useful for images with significant contrast issues or those that require a more detailed examination of the full range of tissue intensities. Together, these codes represent a toolkit for veterinary pathologists to enhance the diagnostic quality of histological images, tailored to the specific needs of each case.

4.3.3 Adaptive Histogram Equalization

The second code snippet that can be inspected in Python Implementation 4.7, builds upon the concepts demonstrated in the first code, offering a more advanced approach to enhancing the visibility of histological images, which is crucial in veterinary pathology. While the first code (Python Implementation 4.6) applies a basic contrast enhancement by scaling the pixel intensity values using the *convertScaleAbs* function (*OpenCV*), this second code introduces a technique known as *contrast limited adaptive histogram equalization* (CLAHE). This method is particularly powerful for enhancing contrast in images where traditional histogram equalization might lead to over-amplification of noise, especially in homogeneous areas like background regions. In this second code (Python Implementation 4.8), the histological image is first loaded and converted from BGR to RGB format, similar to the initial steps in the previous code. However, instead of directly manipulating the contrast through a simple linear transformation, the image is further converted to the LAB color space. The LAB color space separates the lightness component (L) from the color components (A and B), which allows the CLAHE algorithm to focus specifically on enhancing the brightness and contrast of the image without affecting the colors. This is particularly important in histological images, where maintaining accurate color representation is crucial for correct interpretation of tissue structures and patterns.

Adaptive histogram equalization

```
import cv2
from matplotlib import pyplot as plt

# Load the histological
# image from the main path:

image = cv2.imread(r'C:\Users\Elitebook\Desktop\his_img.jpg')
image_rgb = cv2.cvtColor(image, cv2.COLOR_BGR2RGB)

# Convert image to
# LAB color space:

image_lab = cv2.cvtColor(image_rgb, cv2.COLOR_RGB2LAB)

# Apply CLAHE to
# the L channel
# (Lightness):

clahe = cv2.createCLAHE(clipLimit=2.0, tileGridSize=(8, 8))
image_lab[:, :, 0] = clahe.apply(image_lab[:, :, 0])

# Convert back
# to RGB format:

clahe_image = cv2.cvtColor(image_lab, cv2.COLOR_LAB2RGB)

# Display the original
# and CLAHE-enhanced
# images:

plt.figure(figsize=(10, 5))

plt.subplot(1, 2, 1)
plt.title('Original Image')
plt.imshow(image_rgb)

plt.subplot(1, 2, 2)
plt.title('CLAHE Enhanced Image')
plt.imshow(clahe_image)

plt.show()
```

Python implementation 7: Adaptive histogram equalization.

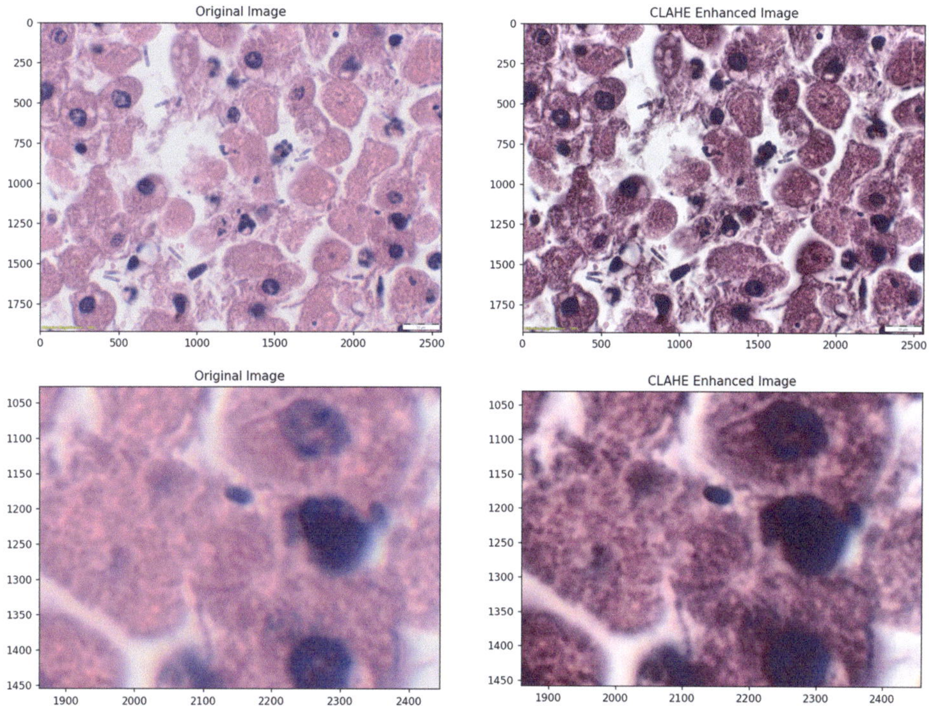

Fig. 4.7 Adaptive histogram equalization on a liver parenchyma of a pig infected with *Clostridium spp*. Images were captured at 100× magnification, with a scale bar representing 10 micrometers. The top-left panel shows the original histological image, revealing the hepatic parenchyma and bacterial infection. The top-right panel presents the processed image using adaptive histogram equalization (AHE), enhancing the contrast to better visualize the structural details and pathological changes. The bottom-left panel provides a zoomed-in view of the original image, focusing on the cellular features and areas of infection. The bottom-right panel shows a zoomed-in view of the processed image, demonstrating improved contrast and detail resolution achieved with AHE

CLAHE is then applied to the lightness channel of the image (the L channel), where it enhances the contrast by equalizing the histogram of small tiles of the image rather than the whole image. This localized approach prevents the over-amplification of noise that can occur with global histogram equalization, making CLAHE especially effective for images with varying levels of brightness and contrast, such as those often encountered in histology slides. The parameters of the algorithm, *clipLimit* and *tileGridSize*, control the extent of the contrast enhancement and the size of the regions used for local histogram equalization, allowing fine-tuning to achieve the desired level of enhancement. After the CLAHE enhancement is applied to the L channel, the image is converted back to the RGB color space for display. The code then uses *Matplotlib* to display the original and CLAHE-enhanced images side by side (Fig. 4.7), allowing for a direct comparison of the effects of the enhancement. In contrast to the first code, where the contrast enhancement was applied uniformly across the entire image (Fig. 4.7), this approach provides a more nuanced enhancement that can reveal subtle details without introducing excessive noise or artifacts.

From a veterinary perspective, Python Implementation 4.7 represents a more sophisticated method for improving the diagnostic quality of histological images. CLAHE is particularly valuable in veterinary pathology when analyzing images with uneven illumination or where certain areas of the tissue might be underexposed. CLAHE can make it easier to detect and evaluate pathological changes in tissues by a simple enhancement of local contrast, such as slight variations in cell density or minor morphological abnormalities that might be missed with simpler contrast enhancement techniques (Fig. 4.7). This can lead to more accurate diagnoses and better-informed treatment decisions in veterinary medicine, particularly in cases where early detection of disease is critical.

4.3.4 Histogram Equalization Using Skimage

The third Python code snippet (Python Implementation 4.8), like the previous examples, is focused on enhancing the contrast of histological images, which is essential for improving the visibility of critical features in veterinary pathology. However, it introduces a different approach by utilizing the *skimage* library, specifically its exposure module, to perform histogram equalization on the image (Fig. 4.8). While the previous codes employed *OpenCV* for image processing—first through a simple contrast enhancement and then using the more advanced CLAHE technique—this code leverages the histogram equalization capabilities of skimage, a library known for its powerful image processing tools tailored for scientific and medical applications. The process begins by loading the histological image using *skimage.io.imread* and converting it to a floating-point representation with *img_as_float*.

Histogram equalization using skimage

```
from skimage import io, exposure, img_as_float
from matplotlib import pyplot as plt

image = img_as_float(io.imread(r'C:\Users\Elitebook\Desktop\his_img.jpg'))

# Apply Histogram Equalization
# using skimage:
hist_eq_image = exposure.equalize_hist(image)

# Display the original and
# histogram-equalized images:

plt.figure(figsize=(10, 5))

plt.subplot(1, 2, 1)
plt.title('Original Image')
plt.imshow(image)

plt.subplot(1, 2, 2)
plt.title('Histogram Equalized Image')
plt.imshow(hist_eq_image)

plt.show()
```

Python implementation 8: Histogram equalization using skimage.

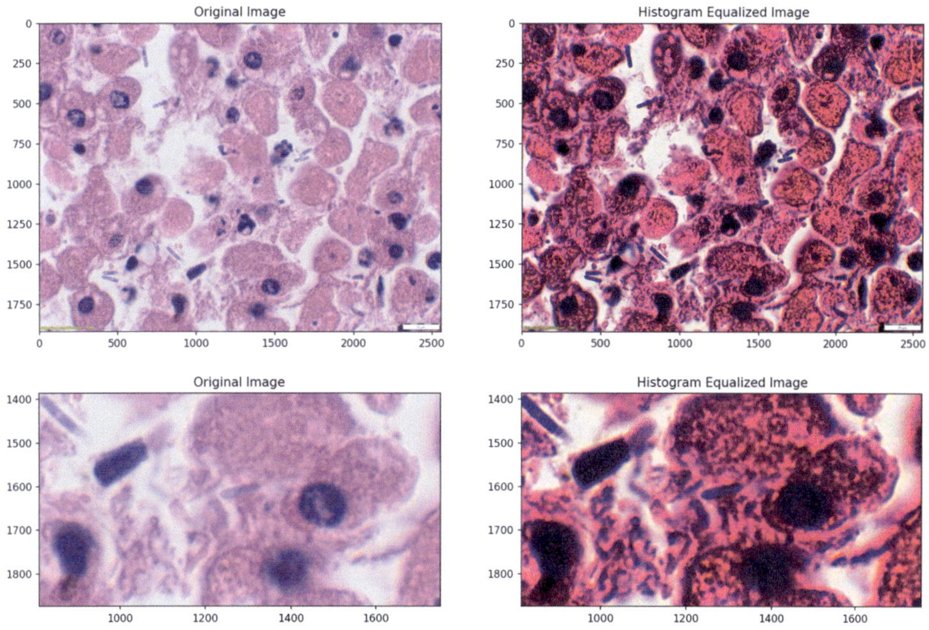

Fig. 4.8 *Skimage* histogram equalization made on a liver parenchyma of a pig with *Clostridium spp*. infection. Images were taken at 100× magnification, with a scale bar representing 10 micrometers. The top-left panel displays the original histological image, highlighting the hepatic parenchyma and areas affected by the bacterial infection. The top-right panel shows the image processed with histogram equalization using skimage, enhancing contrast and making the infected regions and cellular structures more distinct. The bottom-left panel provides a zoomed-in view of the original image, focusing on the finer details of the parenchyma and infected areas. The bottom-right panel presents a zoomed-in view of the processed image, illustrating the enhanced contrast and clarity achieved through histogram equalization, which aids in better visualization of the pathological changes

This step ensures that the image is in a suitable format for the *skimage* operations that follow, similar to how the previous examples involved converting images into formats appropriate for their respective processing methods. Once the image is loaded and converted, the code applies global histogram equalization using *exposure.equalize_hist* from *skimage*. Unlike the localized contrast enhancement provided by CLAHE in the second code, this method spreads out the intensity values of the entire image to enhance its contrast globally. The result is a more uniform distribution of pixel intensities, which can significantly improve the visibility of features across the entire image, especially in histological samples where certain structures may be faint or obscured due to poor contrast (Fig. 4.8). This method of *histogram equalization* is simpler and less computationally intensive compared to CLAHE, making it a quicker option for enhancing contrast when processing time is a consideration.

However, because it applies the same transformation across the entire image, it may not be as effective in dealing with images that have regions of varying contrast. This limitation contrasts with the localized enhancement of CLAHE, which can adjust contrast differently in different parts of the image to avoid amplifying noise or losing detail. In the context of veterinary pathology, this approach can be particularly useful for quickly improving the overall contrast of an image, making it easier to detect broad patterns or features that might otherwise be missed in an unprocessed image. For example, it can enhance the visibility of large-scale structures like tissue boundaries or clusters of cells, which are critical in identifying pathological changes (Fig. 4.8). However, for more detailed analysis where preserving local detail and avoiding noise amplification is important, the CLAHE method from the previous code might be preferable (Python Implementation 4.7). Please note that the code uses *Matplotlib* to display the original and histogram-equalized images side by side (Fig. 4.8). This comparison highlights the impact of the global histogram equalization process on the image, allowing veterinary professionals to assess whether the enhanced image provides better clarity for diagnostic purposes.

4.4 Image Normalization and Calibration

Image normalization and calibration are crucial preprocessing steps in image analysis to ensure that images are standardized and consistent, enabling accurate comparison and analysis. *Image normalization* involves adjusting the pixel intensity values of an image to a common scale, which is important when dealing with images acquired under different conditions. This process typically includes scaling pixel values to a specific range, such as 0 to 1 or 0 to 255, or adjusting the histogram of the image to match a reference distribution. Normalization helps in reducing the effects of lighting variations, contrast differences, and other inconsistencies across images, making the subsequent analysis more robust. Also, *image calibration* refers to the process of correcting systematic errors in an image that may arise due to the imaging system, such as lens distortion, uneven illumination, or differences in magnification. Calibration often involves the use of calibration targets, known reference objects, or mathematical models to adjust the image data so that it accurately reflects the true properties of the sample being imaged. For example, spatial calibration may involve converting pixel measurements into real-world units (e.g., micrometers per pixel), ensuring that measurements and analyses are precise and reliable. Together, *normalization* and *calibration* ensure that images are comparable and that the analysis results are accurate, allowing for consistent interpretation across different images and experimental conditions.

4.4.1 Image Normalization

The code that can be observed in the Python Implementation 4.9 is designed to process and visualize a histological image, particularly useful in the context of veterinary pathology. It utilizes, as before, the *OpenCV* library for image processing and *Matplotlib* for visualization. The main goal of the code is to demonstrate *image normalization*, a crucial step in preparing images for further analysis. First, the code loads a histological image from a specified file path on the computer. This image is then converted to grayscale, simplifying the image data by reducing it to a single channel representing intensity levels. *Grayscale* conversion is often necessary for many image processing tasks, including normalization. Normalization is then applied to the grayscale image (Python Implementation 4.9). This process adjusts the pixel intensity values so that they range from 0 to 1, making the image data consistent and comparable with other images processed in the same way.

Image normalization

```
import cv2
import numpy as np
import matplotlib.pyplot as plt

# Load the image using
# the provided file path:
image_path = r'C:\Users\Elitebook\Desktop\his_img.jpg'
image = cv2.imread(image_path)

# Convert the image to
# grayscale for normalization:
gray_image = cv2.cvtColor(image, cv2.COLOR_BGR2GRAY)

# Normalize the image to have
# pixel values between 0 and 1:
normalized_image = cv2.normalize(gray_image, None, 0, 1, cv2.NORM_MINMAX, dtype=cv2.CV_32F)

# Convert normalized image
# back to 8-bit for visualization:
normalized_image_8bit = (normalized_image * 255).astype(np.uint8)

# Plotting the original and
# normalized images side by
# side:
plt.figure(figsize=(16, 8))

# Original Image:
plt.subplot(1, 2, 1)
plt.imshow(cv2.cvtColor(image, cv2.COLOR_BGR2RGB))
plt.title('Original Image')
plt.axis('on')

# Normalized Image:
plt.subplot(1, 2, 2)
plt.imshow(normalized_image_8bit, cmap='gray')
plt.title('Normalized Image')
plt.axis('on')

# Show window:
plt.show()
```

Python implementation 9: Image normalization.

This step is essential in image analysis, especially when dealing with images captured under varying lighting conditions or with **different equipment**, as it enhances the

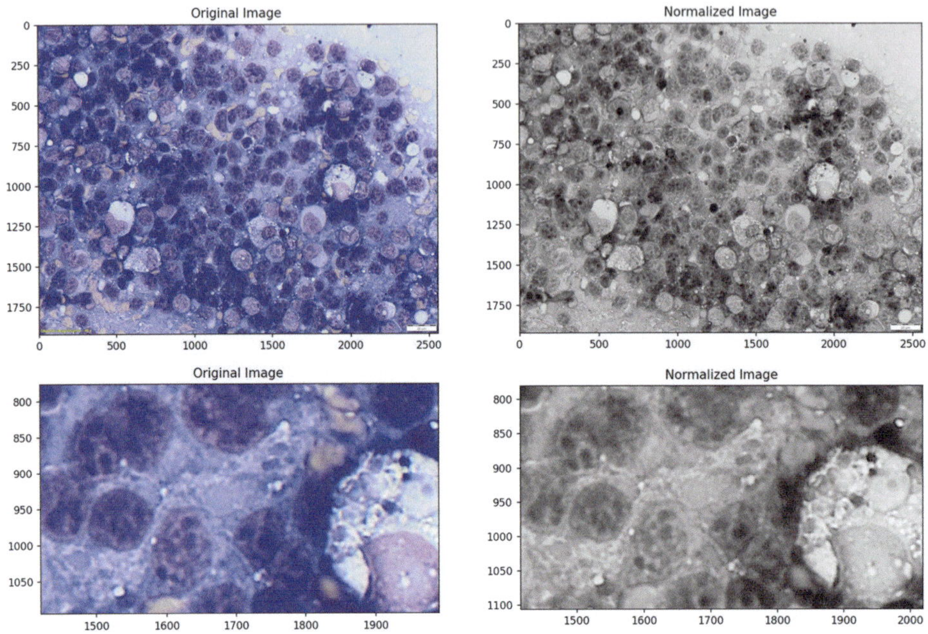

Fig. 4.9 Image normalization experiment made on a cytological sample showing tumoral cells, captured at 40× magnification, with a scale bar representing 20 micrometers. The top-left panel shows the original cytological image, displaying a variety of cellular morphologies and staining intensities (May-Grünwald Giemsa stain). The top-right panel presents the image processed with image normalization, which adjusts the intensity values to a consistent range, enhancing the visibility of subtle features and differences between cells. The bottom-left panel provides a zoomed-in view of the original image, focusing on cellular details such as nuclear and cytoplasmic features. The bottom-right panel shows a zoomed-in view of the processed image, demonstrating how normalization improves the contrast and detail, allowing for more accurate identification and characterization of different cell types

accuracy of any subsequent analysis by ensuring **uniformity** in image data. After normalization, Python Implementation 4.9 converts the normalized image back to an 8-bit format, where pixel values range from 0 to 255. This conversion is necessary for visualization purposes, as most display functions and image file formats expect 8-bit pixel values. The normalized image is then visualized alongside the original image in a side-by-side comparison using *Matplotlib* (Fig. 4.9). This allows the user to visually inspect the effects of normalization, ensuring that the process has been applied correctly and understanding how the image has been adjusted. In veterinary pathology, this code can be useful. Pathologists often work with large sets of histological images, where consistent and accurate image analysis is critical for diagnosing diseases, understanding tissue structure, and conducting research.

This source code (Python Implementation 4.9) helps ensure that subsequent image processing steps, such as segmentation or feature extraction, are based on standardized data, leading to more reliable and reproducible results. This is particularly important when images are used to quantify **cell counts** (please see the following chapters), **measure tissue areas**, or **detect abnormalities**, as any variability in image quality can lead to significant errors in these measurements. Furthermore, the ability to visualize the original and processed images side by side may enable pathologists and researchers to confirm that the preprocessing steps have been correctly applied, further enhancing the accuracy of their analyses.

4.4.2 Image Calibration

The code found in Python Implementation 4.10 performs image calibration and visualization, which is highly relevant in veterinary pathology for ensuring that measurements derived from microscopic images are accurate and standardized. Python Implementation 4.10 uses the *OpenCV* library for image processing and *Matplotlib* for visualization, just as in the examples above. The process begins with loading a histological image from the same specified file path, where the *his_img.jpg* file is found. This image can be a microscopic image of tissue, cells, or any other biological sample relevant to veterinary diagnostics. Once the image is loaded, the code assumes that there is a known reference scale within the image, such as a scale bar that indicates a real-world measurement (e.g., 10 micrometers). In this case, the code assumes that 100 pixels in the image correspond to 10 micrometers in the real world. Using this information, the code calculates the scale (Python Implementation 4.10), specifically the number of micrometers per pixel. This calculation is crucial for converting measurements taken in pixels to actual physical units, allowing for accurate analysis of tissue structures, cell sizes, or other relevant features within the image (Fig. 4.9).

Image calibration

```
import cv2
import numpy as np
import matplotlib.pyplot as plt

# Load the image:
image_path = r'C:\Users\Elitebook\Desktop\his_img.jpg'
image = cv2.imread(image_path)

# Image Calibration: Assuming a known
# scale bar of 10 micrometers equals
# 100 pixels:
known_distance = 10    # in micrometers.
measured_pixels = 100  # in pixels.

# Calculate the scale: micrometers per pixel:
scale = known_distance / measured_pixels

# Annotate the image with the calibration scale:
annotated_image = image.copy()
cv2.putText(
    annotated_image,
    f'Scale: {scale:.2f} um/pixel',
    (10, 30),
    cv2.FONT_HERSHEY_SIMPLEX,
    1,
    (255, 255, 255),
    2,
    cv2.LINE_AA
)

# Add a scale bar to the image:
bar_length = measured_pixels  # Length of the scale bar in pixels.
bar_height = 20               # Height of the scale bar.
bar_color = (255, 255, 255)   # White color for the scale bar.
cv2.rectangle(
    annotated_image,
    (10, annotated_image.shape[0] - 50),
    (10 + bar_length, annotated_image.shape[0] - 50 + bar_height),
    bar_color,
    -1
)
cv2.putText(
    annotated_image,
    f'{known_distance} um',
    (10, annotated_image.shape[0] - 60),
    cv2.FONT_HERSHEY_SIMPLEX,
    1,
    (255, 255, 255),
    2,
    cv2.LINE_AA
)

# Plotting the original and annotated images side by side:
plt.figure(figsize=(12, 6))

# Original Image:
plt.subplot(1, 2, 1)
plt.imshow(cv2.cvtColor(image, cv2.COLOR_BGR2RGB))
plt.title('Original Image')
plt.axis('on')

# Annotated Image with Calibration and Scale Bar:
plt.subplot(1, 2, 2)
plt.imshow(cv2.cvtColor(annotated_image, cv2.COLOR_BGR2RGB))
plt.title('Annotated Image with Calibration and Scale Bar')
plt.axis('on')

# Show the images:
plt.show()
```

Python implementation 10: Image calibration.

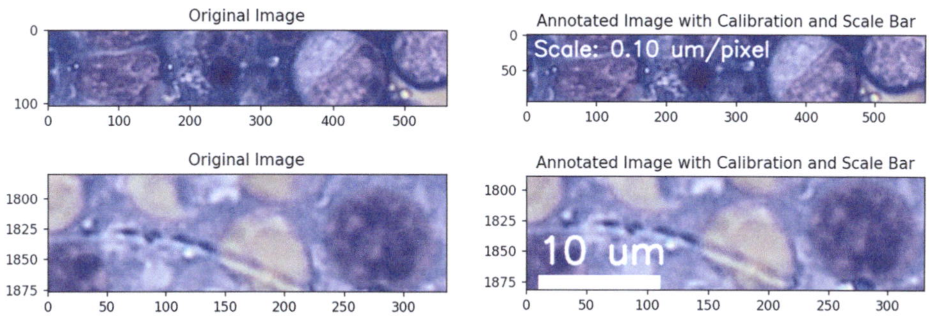

Fig. 4.10 Image calibration made on a cytological sample showing tumoral cells, captured at 40× magnification, with a scale bar representing 20 micrometers. The top-left panel shows the original cytological image, displaying various cellular morphologies. The top-right panel presents the annotated image after processing with image calibration. This annotation includes an overlay displaying the scale (e.g., "Scale: 0.10 um/pixel"), which clarifies how pixel measurements correspond to micrometers. A visual scale bar is also added to the image, representing a known distance of 10 micrometers, positioned in a way that does not obstruct significant cellular details. The scale bar is accurately calibrated to match the measured pixel distance. The bottom-left panel provides a zoomed-in view of the original image, emphasizing cellular structures and their spatial relationships. The bottom-right panel displays a zoomed-in view of the calibrated and annotated image, showing the enhanced clarity of scale representation and spatial measurement, which is critical for precise morphometric analyses

Once the scale is calculated, the code from Python Implementation 4.10 annotates the image with this calibration information (Fig. 4.10). Next, it overlays text on the image that displays the scale (e.g., "Scale: 0.10 um/pixel"), making it clear how pixel measurements translate to micrometers. Additionally, the code adds a visual scale bar to the image. This scale bar is a graphical representation of the known distance (e.g., 10 micrometers) and is placed on the image in a position where it will not obstruct important details. The scale size of the bar in pixels corresponds to the measured distance in micrometers, further reinforcing the calibration.

Next, the code visualizes the results by displaying the original image and the annotated image side by side using *Matplotlib* (Fig. 4.10). This allows the user to compare the unmodified image with the calibrated version, ensuring that the calibration has been applied correctly and that the scale is appropriately represented. Python Implementation 4.10 can be important in veterinary pathology, as this code ensures that measurements made from images are accurate and reproducible. For instance, when assessing tissue samples for signs of disease, pathologists may need to measure the size of cells, the thickness of tissue layers, or the distance between different structures. Accurate calibration allows these measurements to be made in micrometers, which is essential for comparing findings across different cases or studies.

4.5 Image Resizing and Cropping

Image *resizing* and *cropping* are fundamental techniques in image preprocessing, particularly important in fields like veterinary pathology where consistent image dimensions are crucial for accurate analysis and comparison. Image resizing involves changing the dimensions of an image, either by enlarging or reducing its size. This is often necessary when images of different resolutions need to be standardized, for example, when combining images from different sources into a single dataset for analysis. Resizing ensures that all images have the same dimensions, which is particularly important when feeding images into *machine learning* models or performing quantitative analysis. The resizing process typically maintains the aspect ratio of the image to avoid distortion, although in some cases, the aspect ratio may be intentionally altered to fit specific requirements. *Cropping*, on the other hand, involves selecting a specific region of an image and removing the surrounding areas. This technique is used to focus on the region of interest (ROI) within an image, eliminating extraneous details that might distract from the analysis. In veterinary pathology, cropping might be used to isolate a particular tissue section, cell cluster, or pathological feature that needs closer examination. Together, *resizing* and *cropping* are powerful tools for preparing images for further analysis. Moreover, in veterinary pathology, where high-resolution images often contain large amounts of data, these techniques help to optimize image handling and processing. Resizing makes large datasets more manageable by reducing file sizes without losing essential details, while cropping allows for precise examination of specific areas, improving the clarity and relevance of the analysis.

4.5.1 Image Cropping

Building upon the previous discussion of cropping in the context of veterinary pathology, this specific code snippet exemplifies (Python Implementation 4.11) how to extract a focused region from a histological image, enabling detailed analysis of a particular area of interest. In this instance, the coordinates (100, 100) to (500, 500) are used to define a rectangular section of the image, effectively isolating a 400x400 pixel region (Fig. 4.11). This approach allows for precise examination of a specific part of the tissue sample, which might be crucial when dealing with large images where the pathology is localized or when the sample contains multiple areas of interest that need to be examined separately. *This targeted cropping is particularly valuable in cases where the original image contains both healthy and potentially diseased tissue.* The code from Python Implementation 4.11 may enable the pathologist to focus on the pathological features without the distraction of surrounding healthy tissue (by simply isolating only the suspected section). This is critical when assessing fine details such as cellular morphology, tissue architecture, or the presence of specific markers that might be indicative of a disease process.

Image cropping

```
import cv2

# Load the image using
# the provided file path:
image_path = r'C:\Users\Elitebook\Desktop\his_img.jpg'
image = cv2.imread(image_path)

# Define the coordinates for the
# top-left and bottom-right corners
# of the cropping rectangle:
x_start = 450
y_start = 50
x_end = 900
y_end = 500

# Crop the image
cropped_image = image[y_start:y_end, x_start:x_end]

# Save or display
# the cropped image:
cv2.imwrite(r'C:\Users\Elitebook\Desktop\cropped_his_img.jpg', cropped_image)
cv2.imshow('Cropped Image', cropped_image)
cv2.waitKey(0)
cv2.destroyAllWindows()
```

Python implementation 11: Image cropping.

Moreover, by saving the cropped image to a specified file path, this process facilitates the organization and documentation of relevant findings. For example, cropped images can be cataloged and referenced in patient records or research projects, thus ensuring that the most critical aspects of the tissue sample are easily accessible for future review or comparison. Additionally, in veterinary pathology, where time and accuracy are of the essence, this cropping approach can significantly streamline workflows. Pathologists may quickly generate a series of cropped images focusing on different aspects of a sample, which can then be reviewed more efficiently than examining the entire image each time.

This can also aid in discussions with colleagues or in educational settings, where focused images can be used to illustrate specific points or to train students and residents on recognizing key pathological features. Furthermore, the ability to immediately display the cropped image for review ensures that the region of interest has been correctly identified and extracted. This immediate feedback loop may allow the pathologist to adjust the cropping parameters as needed, ensuring that the most relevant and diagnostically useful portions of the image are retained. This level of precision and control is essential in the detailed and careful work that veterinary pathology often requires.

4.5.2 Image Resizing

One fundamental step in pathology workflows is image resizing, which can facilitate the handling, visualization, and analysis of large high-resolution images. In order to continue these examples, the next implementation (Python Implementation 4.12) uses the *OpenCV* library to resize histological images and display them on the screen. It also saves the

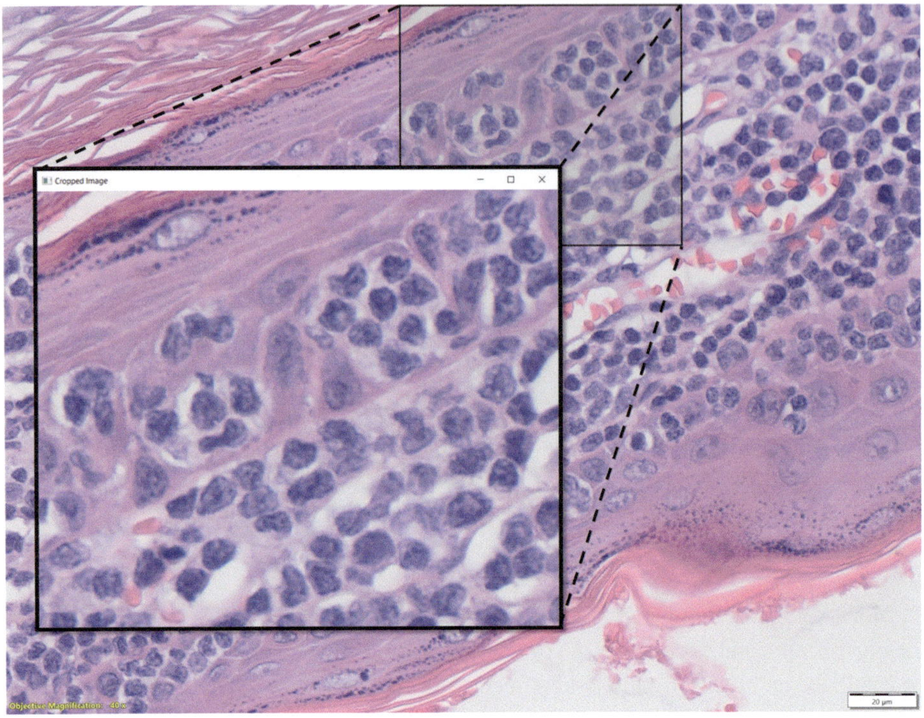

Fig. 4.11 Image cropping on dog skin sample with cutaneous epitheliotropic T cell lymphoma. The image was captured at 40× magnification with a hematoxylin-eosin stain, and the scale bar represents 20 microns. The window highlights the region cropped from the original histological image, focusing on the area of interest where the neoplastic cells clearly infiltrates the epidermis (bottom of image) and adnexal structures, particularly the follicular epithelium (top-left of image). The displayed image not only illustrates the specific features of epitheliotropic lymphoma and intraepidermal clusters of lymphocytes (Pautriers microabscesses) but is also saved to disk for further analysis and documentation, showcasing the precise localization of the pathological changes within the skin tissue

resized image to disk. Resizing is often necessary when working with high-resolution microscopy images because large images can be challenging to manage and analyze. One can more easily visualize specific regions of interest by reducing the image size, improve computational efficiency (smaller size means less information to process), and facilitate faster processing in machine learning or image analysis algorithms.

Image resizing

```
import cv2
image_path = r'C:\Users\Elitebook\Desktop\his_img.jpg'
image = cv2.imread(image_path)

# Check if the image is loaded:
if image is None:
    print("Error loading image. Please check the image path and file integrity.")
else:

    # Resize the image to a new size
    # (e.g., 40% of the original size):
    width = int(image.shape[1] * 0.4)   # 40% of the original width.
    height = int(image.shape[0] * 0.4)  # 40% of the original height.

    resized_image = cv2.resize(image, (width, height), interpolation=cv2.INTER_AREA)

    # Display the resized
    # image using OpenCV:

    cv2.imshow('Resized Image', resized_image)

    # Wait for any key press
    # to close the window:

    cv2.waitKey(0)
    cv2.destroyAllWindows()

    # Save the resized
    # image to disk:

    save_path = r'C:\Users\Elitebook\Desktop\resized_his_img.jpg'
    cv2.imwrite(save_path, resized_image)  #
    print(f"Resized image saved to: {save_path}")
```

Python implementation 12: Image resizing.

The code begins by specifying the file path of a histological image in order to load it using *cv2.imread*(), a function from the *OpenCV* library that reads the image file into a variable named image. It then checks whether the image has been loaded correctly. If the image is not loaded (e.g., due to an incorrect path or a corrupted file), an error message is printed. Once the image is loaded successfully, the code calculates the new dimensions for resizing the image to 40% of its original size. The resizing is done using *cv2.resize()*, which is a highly efficient function that resamples the image according to the specified dimensions. In this case, *cv2.INTER_AREA* interpolation is used, which is suitable for downscaling and helps preserve the quality of the image by averaging pixel values. Once resizing is done, the script displays the resized image in a new window using *cv2.imshow()*. The user can visually inspect the resized image to ensure that it retains the essential features necessary for diagnosis or further analysis (Fig. 4.12).

The window remains open until the user presses a key, after which it closes automatically (the use of *cv2.waitKey(0)* method). After the inspection, *cv2.destroyAllWindows()* closes the display window (Fig. 4.12). Next, the resized image is saved to disk at a specified path using *cv2.imwrite()*. This step ensures that the processed image can be reused for documentation, sharing, or further computational analysis, such as feature extraction or classification using machine learning models. Please note that this code provides a straightforward yet effective approach to manage and preprocess histological images for

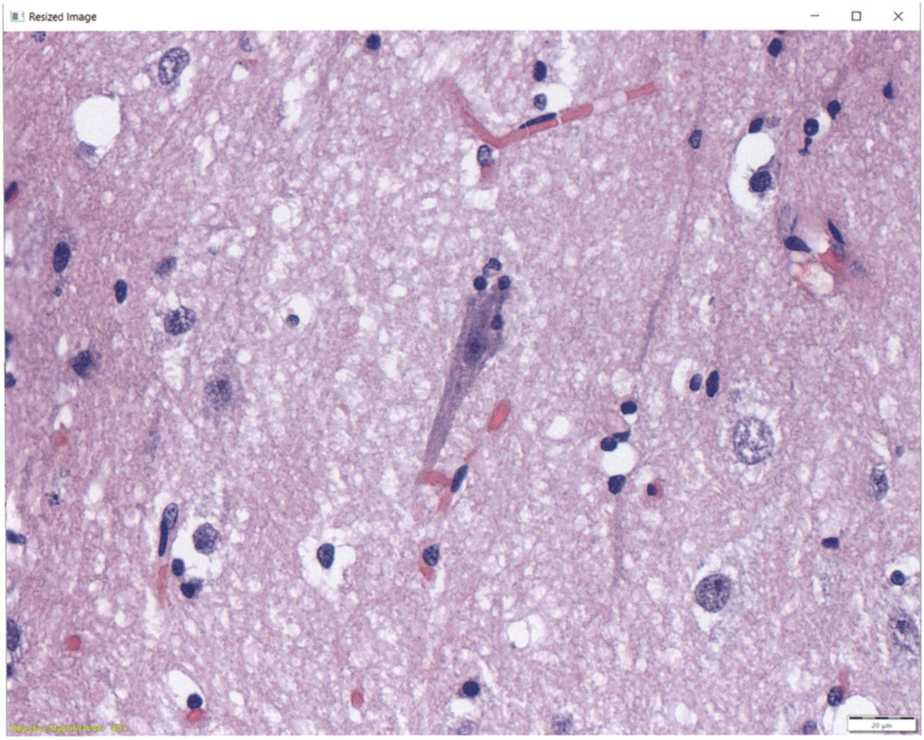

Fig. 4.12 Image resizing on brain tissue of a dog showing neuronal degeneration, satellitosis, astrogliosis, and spongiosis (Aujeszky's disease or pseudorabies). The image was captured at 40× magnification, with a scale bar representing 20 micrometers. The displayed window provides the size of the processed image and also indicates that the program saves the processed file to disk. The histological features of interest, such as neuronal satellitosis (clustering of microglial cells around degenerated neurons), astrogliosis (proliferation of astrocytes and reactive astrocytes, also known as astrocytosis), and spongiosis (vacuolization within the neutrophil), are highlighted for detailed examination and analysis

diagnostic purposes. Resized images can be integrated into a digital pathology pipeline, aiding in the development of AI-based diagnostic tools, thus improving data management and enhancing the efficiency of pathological assessments.

4.6 Data Augmentation for Microscopy Images

In the context of veterinary pathology, where obtaining large datasets can be challenging due to the labor-intensive process of collecting and annotating high-quality images, data augmentation serves as an essential tool to improve model robustness and performance [113]. *Data augmentation* for microscopy images is a powerful technique used to enhance the diversity of image data available for training machine learning models. Data augmentation involves artificially expanding the size of a dataset by applying various transformations to the original images. These transformations can include operations such as *rotations*, *flips*, *translations*, *zooms*, and even adjustments to *brightness* and *contrast*. When multiple variations are generated for each image, *data augmentation* helps to simulate the variability that might be encountered in real-world scenarios, ensuring that the model learns to generalize better across different types of images. For instance, in veterinary pathology, tissue samples might be imaged from slightly different angles or under varying lighting conditions. When applying rotations and brightness adjustments as part of *data augmentation*, the model can learn to recognize pathological features regardless of these variations. This is particularly useful when the goal is to develop a model that can reliably detect diseases across different samples and imaging conditions. Moreover, data augmentation can help address the issue of class imbalance, which is common in veterinary pathology datasets where some pathological conditions may be underrepresented [114]. Apparently, only by augmenting images of the minority class more aggressively, the dataset can be balanced, reducing the risk of the model becoming biased toward the more prevalent conditions [114]. Another important aspect of data augmentation in microscopy is its ability to introduce subtle variations that are often present in real biological samples. For example, slight changes in the orientation of cells or variations in staining intensity can be simulated through augmentation. This ensures that the model does not overfit to specific patterns present in the training data but instead learns the underlying features that distinguish different types of cells or tissues. In practice, data augmentation can be implemented during the training phase of a machine learning model, where each image in the dataset is randomly transformed before being fed into the model. This on-the-fly augmentation means that the model never sees the exact same image twice, significantly enhancing its ability to generalize.

4.6.1 A Concrete Example Data Augmentation

As stated above, data augmentation for microscopy images in veterinary pathology is not just a method for increasing dataset size but a critical technique for improving model accuracy, generalization, and robustness. Simulation of variability and complexity of real-world samples allows for data augmentation which ensures that machine learning models are well equipped to handle the variations and sometimes unpredictable nature of veterinary pathology images, leading to more reliable diagnostic tools and research outcomes. The new example found in Python Implementation 4.13 uses *Keras API* of *TensorFlow* to perform data augmentation on microscopy images, which is a critical preprocessing step in preparing image datasets for machine learning models, particularly in fields like veterinary pathology. The code starts by loading the input image like before, by using *OpenCV*, and converting it to RGB format, which is required for proper processing by *Keras*. The image is then reshaped to include a batch size dimension, which *Keras* expects when handling image data. This reshaping is necessary because *Keras* operates on batches of images rather than single images, even when only one image is being processed.

Data augmentation

```
import tensorflow as tf
from tensorflow.keras.preprocessing.image import ImageDataGenerator
import cv2
import matplotlib.pyplot as plt

# Load the image using
# the provided file path.

image_path = r'C:\Users\Elitebook\Desktop\his_img.jpg'

image = cv2.imread(image_path)

# Convert to RGB for Keras:

image = cv2.cvtColor(image, cv2.COLOR_BGR2RGB)

# Reshape the image to include
# batch size (Keras expects a
# 4D array: batch_size, height,
# width, channels).

image = image.reshape((1,) + image.shape)

# Create an instance of the
# ImageDataGenerator with
# various augmentations.

datagen = ImageDataGenerator(
    rotation_range=40,      # Randomly rotate images by up to 40 degrees.
    width_shift_range=0.2,  # Randomly shift images horizontally by up to 20% of the width.
    height_shift_range=0.2, # Randomly shift images vertically by up to 20% of the height.
    shear_range=0.2,        # Shear the images.
    zoom_range=0.2,         # Randomly zoom into images by up to 20%.
    horizontal_flip=True,   # Randomly flip images horizontally.
    fill_mode='nearest'     # Fill in missing pixels after a transformation.
)

# Generate augmented
# images and display them:

plt.figure(figsize=(12, 12))
i = 0
for batch in datagen.flow(image, batch_size=1):
    plt.subplot(3, 3, i + 1)
    augmented_image = batch[0].astype('uint8')
    plt.imshow(augmented_image)
    plt.axis('off')
    i += 1
    if i % 9 == 0:
        break  # Stop after generating 9 augmented images ...

plt.show()
```

Python implementation 13: Data augmentation.

Next, the core of the code involves the *ImageDataGenerator* class from *Keras*, which is configured to apply a series of random transformations to the image (Python Implementation 4.13). These transformations include rotating the image by up to 40 degrees, shifting it horizontally and vertically by up to 20% of its dimensions, *shearing*, *zooming*, and *flipping* the image horizontally (Fig. 4.13). These operations generate new, slightly altered versions of the original image, effectively increasing the variation of the

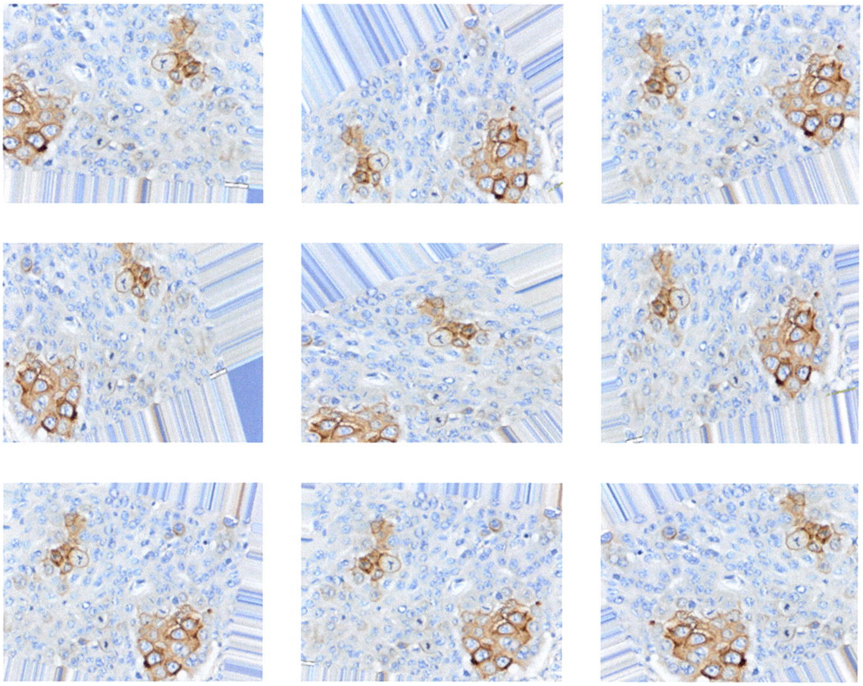

Fig. 4.13 Augmentation made on dog skin sample showing a poorly differentiated malignant tumor. The main image was captured at 40× magnification, with a reference scale bar of 20 micrometers. Immunohistochemistry was performed using the Pan-Cytokeratin antibody (panCK), which is a marker for the differentiation of epithelial cells, and here assesses the presence of cells of epithelial origin. The figure presents a 3×3 matrix structure where each element represents an augmented version of the original image, demonstrating various transformations such as rotations, flips, and intensity changes. These augmented images are potential components of a training dataset for machine learning applications, allowing the model to learn robust features for identifying epithelial differentiation in anaplastic tumors

training dataset without the need for additional manual data collection. The *fill_mode='nearest'* parameter ensures that any pixels generated by these transformations (e.g., when *rotating* or *shifting* the image) are filled in a reasonable way, using the nearest available pixel values. The acquisition of a large and various dataset can be challenging due to the time and resources required for manual image collection and annotation.

Data augmentation addresses this challenge by artificially expanding the dataset. The utilization of these augmented images for training allows the model to become more adept at generalizing to new, unseen images that might vary in orientation, position, or scale, making it more reliable in real-world diagnostic scenarios. Note that to use this code (Python Implementation 4.13), *TensorFlow* must be installed in the Python environment. *TensorFlow* is a powerful open-source platform for machine learning and artificial intelligence. If this code is run for the first time and *TensorFlow* is not already installed, one can install it using the Python package manager *pip*. The installation command is:

Installation command

```
pip install tensorflow
```

Once installed, *TensorFlow* provides the necessary tools to build, train, and deploy machine learning models, including utilities like the *ImageDataGenerator* for *data augmentation*. After installation, this code can be executed to generate augmented images, which can be used to train models that are more resilient to variations in real-world data, ultimately improving the accuracy and reliability of veterinary diagnostic tools. This is especially useful in veterinary pathology, where the variability in biological samples can be high, and robust models are essential for accurate disease detection and classification.

Image Segmentation and Feature Extraction 5

Abstract

This chapter introduces some practical examples for image segmentation and feature extraction as essential processes in computational image analysis for veterinary pathology. Image segmentation divides an image into distinct regions representing meaningful components like cells or tissue structures, a critical step for accurate diagnosis in veterinary pathology. Techniques such as thresholding, edge detection, and advanced machine learning models, including convolutional neural networks (CNNs), are discussed for their ability to delineate pathological features such as tumor boundaries or necrotic areas. Feature extraction, which follows segmentation, focuses on quantifying characteristics of the segmented regions, such as shape, texture, color, and intensity, to facilitate further analysis or diagnosis. The interplay between segmentation and feature extraction enhances the interpretation of complex histological images, providing insights into disease mechanisms, tissue morphology, and cellular behavior. Applications of these methods in veterinary pathology are explored, particularly their roles in automating diagnostics, improving reproducibility, and advancing research. This chapter also highlights the importance of preprocessing steps like noise reduction and contrast enhancement, which improve segmentation accuracy, ultimately leading to more precise and reliable diagnoses.

Keywords

Thresholding · Edge detection · Region-growing · Watershed algorithm · Morphological operations

E. Gagniuc, *Computational Analysis of Microscopy Images*,
https://doi.org/10.1007/978-3-031-86270-0_5

5.1 Introduction

In the field of computational image analysis for veterinary pathology, image segmentation and feature extraction are foundational processes that underpin a wide range of applications, from automated diagnostics to advanced research in tissue morphology. Image segmentation involves partitioning an image into distinct regions that correspond to meaningful components, such as cells, tissue structures, or regions of interest in histological slides. This step is critical because it directly impacts the accuracy of subsequent tasks like feature extraction, object recognition, and classification. In veterinary pathology, precise segmentation is essential for identifying pathological features such as tumor boundaries, areas of necrosis, or inflamed regions. Segmentation techniques range from simple methods like thresholding and edge detection to more sophisticated approaches such as *region-based segmentation*, *morphological operations*, and *machine learning* models like *convolutional neural networks* (CNNs), each with unique strengths and applications. Following segmentation, feature extraction focuses on identifying and quantifying key characteristics of the segmented regions to facilitate further analysis, classification, or diagnosis. Features can include shape, texture, color, intensity, and geometric properties, each providing valuable insights into tissue types or pathological changes. For instance, shape features help distinguish between normal and cancerous cells, texture features analyze tissue patterns indicative of specific diseases, and color features quantify staining differences in histological slides. Effective feature extraction is crucial as it transforms raw image data into meaningful descriptors that can be used for predictive modeling or comparative analysis, aiding in the differentiation between healthy and diseased tissues, cell classification, and even predicting disease outcomes. Image segmentation and feature extraction are interdependent; accurate segmentation is vital for reliable feature extraction, and extracted features can refine segmentation algorithms, particularly in iterative or machine learning contexts. Together, these techniques enable a detailed analysis and interpretation of complex histological images, providing powerful tools for advancing veterinary diagnostics and research.

5.2 Thresholding and Edge Detection

Thresholding and edge detection are fundamental techniques in image processing, particularly useful in preparing images for further analysis by segmenting and identifying important features. Thresholding is a simple yet effective method for segmenting images by converting grayscale images into binary images [115]. This process involves setting a specific threshold value and then categorizing each pixel in the image as either part of the foreground or background based on whether its intensity is above or below this threshold. Common thresholding techniques include *global thresholding*, where a single threshold value is applied to the entire image, and *adaptive thresholding*, where the threshold value varies across different regions of the image to account for varying lighting conditions. *Otsu method* is an *advanced global thresholding* technique that automatically determines the optimal threshold by minimizing intra-class variance [89]. On the other hand, *edge detection* is another crucial technique used to identify boundaries within images. It works by detecting the sharp changes in intensity, which often correspond to object boundaries. The *Sobel operator* is one of the simplest and most commonly used methods, which calculates the gradient of the image intensity at each pixel, providing an approximation of the intensity changes. The *Canny edge* detector is a more advanced and widely used algorithm that improves upon basic edge detection by using multiple stages, including noise reduction with *Gaussian filtering*, finding intensity gradients, non-maximum suppression to thin out the edges, and double thresholding to identify strong and weak edges, followed by edge tracking by hysteresis. These techniques are particularly valuable in medical image analysis, such as in histology, where clear delineation of cellular structures is essential. Please note that these preparatory steps are often the first stage in more complex image analysis workflows, including segmentation and feature extraction.

5.2.1 Thresholding

A practical example is provided for each of the main methods mentioned above. Thus, the code shown in Python Implementation 5.1 uses the *OpenCV* and *Matplotlib* libraries to apply global thresholding to a grayscale image, which is particularly useful in veterinary pathology for isolating specific structures within a tissue sample, such as normal cells or pathological regions. The process begins by loading a histological image in grayscale from a specified file path. Please note that grayscale images are ideal for many image processing tasks because they simplify the data, focusing on intensity values rather than color, which is especially useful when dealing with stained tissue samples commonly used in pathology. The core operation of the code from Python Implementation 5.1 is global thresholding, applied using the *cv2.threshold*() function. In this case, a threshold value of 100 is set, meaning that any pixel in the image with an intensity value above 100 is turned white (150 in this specific example), and any pixel below this threshold is turned black (0). This operation results in a binary image, where key structures of interest are highlighted, and less relevant background information is suppressed.

Thresholding

```
import cv2
import matplotlib.pyplot as plt

# Load the image in grayscale:
image_path = r'C:\Users\Elitebook\Desktop\his_img.jpg'
image = cv2.imread(image_path, cv2.IMREAD_GRAYSCALE)

# Apply global thresholding:
_, thresholded_image = cv2.threshold(image, 100, 150, cv2.THRESH_BINARY)

# Display the original
# and thresholded images.

plt.figure(figsize=(12, 6))

plt.subplot(1, 2, 1)
plt.imshow(image, cmap='gray')
plt.title('Original Image')
plt.axis('on')

plt.subplot(1, 2, 2)
plt.imshow(thresholded_image, cmap='gray')
plt.title('Thresholded Image')
plt.axis('on')

plt.show()
```

Python implementation 14: Thresholding.

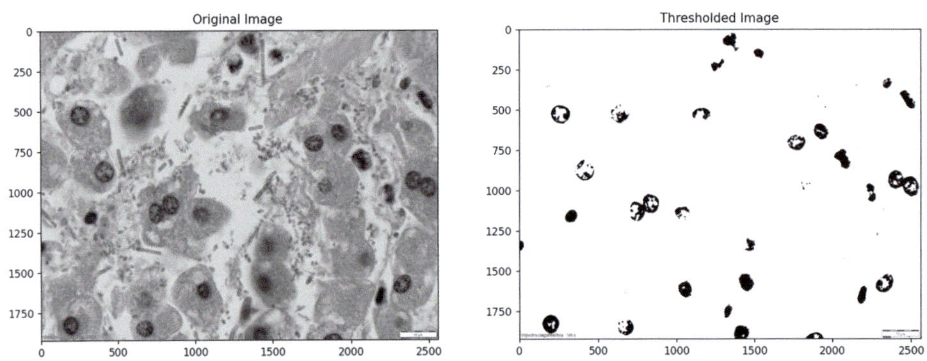

Fig. 5.1 Thresholding on liver parenchyma of a pig infected with *Clostridium* spp. Images were taken at 100× magnification, with a scale bar representing 10 micrometers. The left panel shows the original histological image, displaying the hepatic parenchyma with evident infection. The right panel presents the processed image using a thresholding binary mask, which segments and highlights the infected regions and bacterial presence by converting them into distinct binary areas, facilitating the identification and analysis of pathological changes

This simplification is crucial for tasks where only certain aspects of the image, such as specific cell types or regions with abnormal staining, need to be analyzed. Once the thresholding is applied, the code uses *Matplotlib* to display the original grayscale image alongside the thresholded image (Fig. 5.1). This visual comparison allows for a quick assessment of the thresholding process, ensuring that the selected threshold effectively highlights the regions of interest without losing important information. Python Implementation 5.1 may be valuable in the context of veterinary pathology, because it enables pathologists to quickly isolate and analyze specific features within a tissue sample. For example, thresholding can help differentiate between healthy and diseased tissue areas by highlighting regions with higher or lower staining intensities. This is particularly useful in diagnosing conditions where certain cells or tissues may show characteristic staining patterns, such as in cancerous tissues or inflammatory conditions.

Moreover, thresholding can be a precursor to more advanced image analysis tasks, such as counting cells, measuring tissue areas, or performing automated classification (please see the next chapter). The conversion of a complex histological image to a simpler binary format allows thresholding to facilitate the application of these additional analytical techniques. This not only speeds up the diagnostic process but also increases accuracy by reducing the potential for human error in manual image interpretation. Nevertheless, the code seen in Python Implementation 5.1 provides a powerful and straightforward tool for enhancing the analysis of histological images in veterinary pathology, helping to improve diagnostic accuracy and efficiency. Last but not least, note that genetics research can also be integrated using this thresholding methodology, as the black areas represent heterochromatin in the cell nucleus, which in turn indicates a general status of cellular activity (Fig. 5.1).

5.2.2 Edge Detection

The next code (Python Implementation 4.11) builds upon the concepts introduced in the previous example by shifting the focus from *thresholding* to *edge detection*, specifically using the *Canny edge* detection algorithm. While thresholding is useful for isolating areas of interest based on pixel intensity, edge detection focuses on identifying the boundaries and contours within the image, providing a complementary approach to image analysis in veterinary pathology. The code begins similarly to the previous one by loading a histological image in grayscale. This grayscale conversion is essential because edge detection relies on intensity gradient variations in brightness across the image to identify the edges of structures such as cells or tissue boundaries (Fig. 5.2). Therefore, the key operation in Python Implementation 5.2 is the application of the *Canny edge* detection algorithm via the *cv2.Canny*() function. Here, the function is configured with two threshold values, 30 and 100, to control the sensitivity of edge detection. The lower threshold detects faint edges, while the higher threshold discards weak edges and focuses on the more prominent ones. This dual-threshold approach ensures that only the most significant edges are captured, which is crucial when analyzing complex histological images where fine details must be preserved.

Edge detection

```
import cv2
import matplotlib.pyplot as plt

# Load the image in grayscale.
image_path = r'C:\Users\Elitebook\Desktop\his_img.jpg'
image = cv2.imread(image_path, cv2.IMREAD_GRAYSCALE)

# Apply Canny edge detection
# with even lower thresholds.
edges = cv2.Canny(image, 30, 100)

# Display the original and
# edge-detected images.
plt.figure(figsize=(12, 6))

plt.subplot(1, 2, 1)
plt.imshow(image, cmap='gray')
plt.title('Original Image')
plt.axis('on')

plt.subplot(1, 2, 2)
plt.imshow(edges, cmap='gray')
plt.title('Edge Detected Image')
plt.axis('on')

plt.show()
```

Python implementation 15: Edge detection.

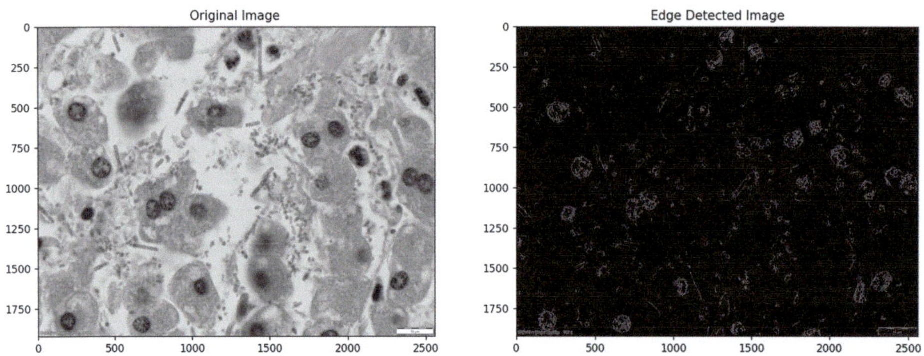

Fig. 5.2 Edge detection on liver parenchyma of a pig infected with *Clostridium* spp. Images were captured at 100× magnification, with a scale bar representing 10 micrometers. The left panel displays the original histological image, highlighting the hepatic parenchyma and areas affected by the bacterial infection. The right panel shows the processed image using edge detection with a binary mask, which accentuates the boundaries of cellular structures and highlights the distribution of bacteria within the infected tissue

When the edge detection is applied, the code shown in Python Implementation 5.2 uses *Matplotlib* (as in the previous examples) to display the original image alongside the edge-detected image (Fig. 5.2). This side-by-side comparison is valuable for assessing how well the edges of structures within the tissue sample have been identified, providing immediate visual feedback on the effectiveness of the edge detection process. In veterinary pathology, the value of this code (Python Implementation 5.2) lies in its ability to delineate the boundaries of various structures within a tissue sample. *Edge* detection is particularly useful for outlining cells, identifying the edges of lesions, or segmenting different tissue regions. A clear definition of these boundaries allows pathologists to more accurately measure and analyze the size, shape, and distribution of cells or pathological features, which are critical factors in diagnosing diseases, such as cancer or inflammatory conditions.

Note that the edge method and the threshold method may exhibit imperceptibly different results for the same input image. Specifically, the *edge* perimeter of the same cell nucleus is slightly larger than the perimeter that can be measured on an image using *thresholding* (please make a visual comparison between Figs. 5.1 and 5.2). This is because the edges of the cell nucleus are less taken into account by the thresholding method, as the perimeter color is outside the imposed threshold. Nevertheless, when used in conjunction with the previous *thresholding* code (Python Implementation 5.1), edge detection offers a more complete image analysis toolkit. *Thresholding* simplifies the image by highlighting areas of interest, while *edge* detection further refines the analysis by focusing on the contours of those areas. Together, these techniques enable more precise and reliable analysis of histological images, improving diagnostic accuracy and aiding in the development of automated tools for veterinary pathology.

5.3 Region-Based Segmentation Techniques

Region-based segmentation techniques divide an image into meaningful regions based on the similarity of pixel values, such as intensity, color, or texture. These methods focus on grouping pixels that share specific properties to form connected regions that correspond to distinct objects or areas within the image. One common approach is *region growing*, which starts from a set of seed points and expands the regions by adding neighboring pixels that meet predefined criteria for similarity (Python Implementation 5.3). This technique is useful for segmenting regions with well-defined boundaries but can be sensitive to noise and the choice of initial seeds. Another widely used method is the *watershed* algorithm, which treats the grayscale image as a topographic surface where bright pixels represent peaks and dark pixels represent valleys (Python Implementation 5.4). This technique finds the watershed lines, or boundaries, by simulating how water would flow over the surface, effectively separating adjacent regions. The watershed algorithm is particularly effective for **separating touching objects**, such as **overlapping cells**, but may require preprocessing steps like noise reduction or markers to prevent over-segmentation. Thus, region-based techniques are valuable in histological image analysis as they enable precise segmentation of tissues, cells, and other structures, facilitating detailed examination and quantification of morphological features.

5.3.1 Region Growing

The following code (Python Implementation 5.3) demonstrates the use of region-based segmentation, specifically through the *flood fill* algorithm, to isolate and analyze regions within a histological image [116]. The process starts by loading the image in grayscale, which simplifies the data by focusing solely on intensity values, making it easier to process for segmentation tasks. To enhance the visibility of the regions of interest, the code shown in Python Implementation 5.3 applies a more aggressive thresholding technique. This converts the grayscale image into a binary image, where the pixels are either black or white (Fig. 5.3). The thresholding process is inverted (i.e., *cv2.THRESH_BINARY_INV*), meaning that areas below the threshold become white, and those above it become black. This binary image serves as the basis for region growing, where the goal is to identify and isolate a specific area of the tissue sample that is visually distinct from the rest.

Region growing

```
import cv2
import numpy as np
import matplotlib.pyplot as plt

# Load the image
# in grayscale:

image_path = r'C:\Users\Elitebook\Desktop\his_img.jpg'
image = cv2.imread(image_path, cv2.IMREAD_GRAYSCALE)

# Apply a more aggressive
# threshold to create a
# more prominent binary
# image:

_, binary_image = cv2.threshold(image, 100, 255, cv2.THRESH_BINARY_INV)

# Choose a different seed
# point (e.g., a point with
# more contrast). Ex. the
# center of the image:

seed_point = (image.shape[1] // 2, image.shape[0] // 2)

# Create a mask for region growing
# (size is 2 pixels larger than
# the image in each dimension):

mask = np.zeros((image.shape[0] + 2, image.shape[1] + 2), np.uint8)

# Perform region growing
# using the floodFill
# function:

_, _, mask, _ = cv2.floodFill(binary_image, mask, seed_point, 255, loDiff=5, upDiff=5,
flags=cv2.FLOODFILL_MASK_ONLY | (4 << 8))

# Display the original
# and region grown
# images:

plt.figure(figsize=(12, 6))

plt.subplot(1, 2, 1)
plt.imshow(image, cmap='gray')
plt.title('Original Image')
plt.axis('on')

plt.subplot(1, 2, 2)

# Remove the extra
# padding added by
# floodFill:

plt.imshow(mask[1:-1, 1:-1], cmap='gray')
plt.title('Region Grown Image')
plt.axis('on')

plt.show()
```

Python implementation 16: Region growing.

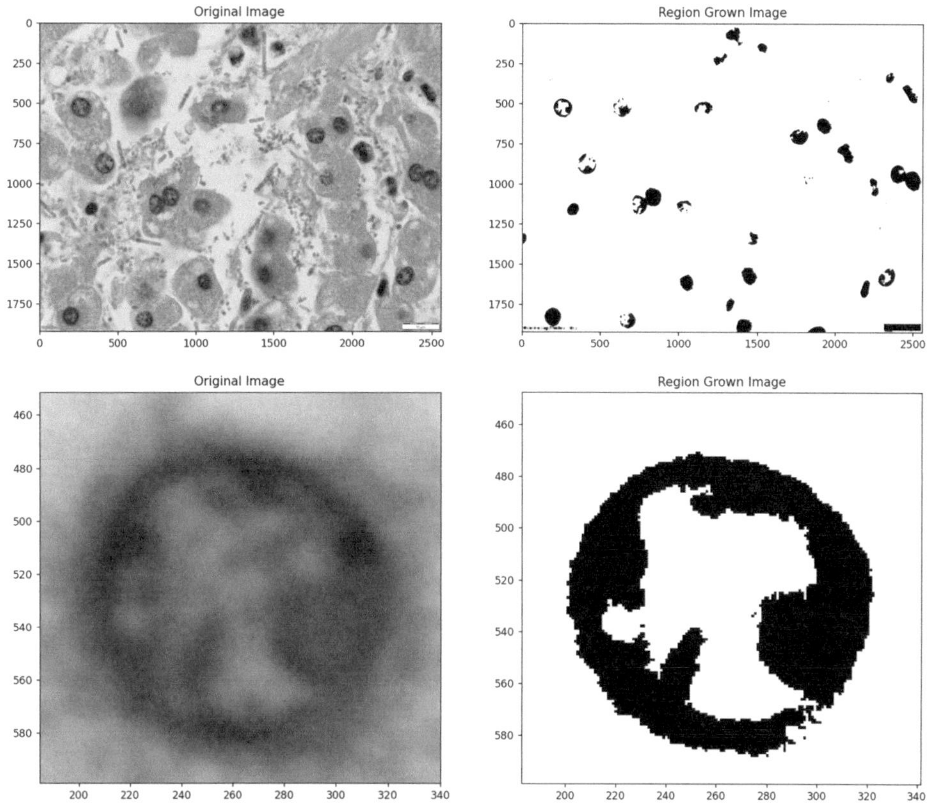

Fig. 5.3 Region growing applied to images showing liver parenchyma of a pig infected with *Clostridium* spp. Images were captured at 100× magnification, with a scale bar representing 10 micrometers. The top-left panel shows the original histological image, highlighting the hepatic tissue and bacterial infection. The top-right panel displays the image after processing with the region growing algorithm, resulting in a binary image that segments and isolates specific regions of interest, such as infected areas. The bottom-left panel provides a zoomed-in view of the original image, focusing on the detailed cellular architecture and the presence of bacteria. The bottom-right panel presents a zoomed-in view of the processed binary image, showing distinct segmented regions that facilitate the visualization and identification of infected zones and tissue changes

As mentioned, the core of the code is the *flood fill* algorithm, which is a region-based segmentation method. The algorithm starts from a seed point, which in this case is chosen as the center of the image. From this seed point, the algorithm "grows" the region by filling neighboring pixels that have similar intensity values, as defined by the *loDiff* and *upDiff* parameters. These parameters control how similar neighboring pixels need to be to the seed point for them to be included in the growing region. The result is a mask that highlights the grown region, effectively segmenting it from the rest of the image. Next, the code displays both the original grayscale image and the mask resulting from the *flood fill* operation (Fig. 5.3). This allows for a visual comparison, showing how the region of interest has been successfully isolated from the rest of the tissue.

This code (Python Implementation 5.3) is particularly valuable in veterinary pathology for isolating and analyzing specific regions within a tissue sample. For example, in the diagnosis of tumors, it is often necessary to isolate the tumor mass from surrounding healthy tissue to assess its size, shape, and boundaries. The flood fill algorithm is well suited for this task because it can precisely define regions that are visually distinct, such as a tumor that appears darker or lighter than the surrounding tissue due to differences in cell density or staining. This technique is also useful in research settings where the morphology of specific tissue regions needs to be studied in detail. For instance, researchers studying inflammatory diseases might use region growing to isolate areas of inflammation within a tissue sample, allowing them to quantify the extent of inflammation and analyze its impact on surrounding tissues. Moreover, such a method is helpful in high-throughput analysis, where large numbers of tissue samples need to be processed quickly and consistently. As noted in the previous method, this approach can also incorporate genetic research, as the black regions correspond to heterochromatin. The extent of heterochromatin coverage over the cell nucleus can serve as an indicator of cell activity level (Fig. 5.3).

5.3.2 Watershed Algorithm

This code from Python Implementation 5.4 represents an advanced image segmentation technique, building upon the concepts introduced in the previous examples of thresholding and region-based segmentation. Specifically, it utilizes the *watershed* algorithm, a powerful method for separating overlapping objects within an image, which is particularly relevant in the complex and intricate field of veterinary pathology [117]. This example begins with the same foundational step as in the previous examples, namely loading a histological image in grayscale. However, this implementation adds several preprocessing steps to enhance the image and prepare it for more accurate segmentation. First, a *Gaussian blur* is applied to reduce noise, which is essential for preventing small, irrelevant details from interfering with the segmentation process. This is followed by thresholding to create a *binary image*, where pixel values are set to either black or white based on intensity. The binary image is then cleaned up using morphological opening, which removes small objects that might be misinterpreted as regions of interest. The code further refines the image by applying a *distance transform*, which converts the binary image into a map where pixel values represent the distance to the nearest background pixel. This transformation is normalized to enhance the contrast between different regions. Thus, a threshold imposed on the distance map allows the code to identify "sure foreground," namely regions that are definitely part of the objects of interest, such as cells or tissue structures.

Watershed algorithm

```
import cv2
import numpy as np
import matplotlib.pyplot as plt

# Load the image in grayscale:
image_path = r'C:\Users\Elitebook\Desktop\his_img.jpg'
image = cv2.imread(image_path, cv2.IMREAD_GRAYSCALE)

# Apply GaussianBlur
# to reduce noise:
blurred_image = cv2.GaussianBlur(image, (5, 5), 0)

# Apply a threshold to
# create a binary image:
_, binary_image = cv2.threshold(blurred_image, 120, 255, cv2.THRESH_BINARY_INV)

# Remove noise with
# morphological opening:
kernel = np.ones((3, 3), np.uint8)
opening = cv2.morphologyEx(binary_image, cv2.MORPH_OPEN, kernel, iterations=3)

# Distance transform:
dist_transform = cv2.distanceTransform(opening, cv2.DIST_L2, 5)

# Normalize distance transform:
dist_transform = cv2.normalize(dist_transform, None, 0, 1.0, cv2.NORM_MINMAX)

# Apply a threshold to
# get sure foreground:
_, sure_fg = cv2.threshold(dist_transform, 0.5 * dist_transform.max(), 255, 0)
sure_fg = np.uint8(sure_fg)

# Get the sure background by
# dilating the original binary
# image:
sure_bg = cv2.dilate(opening, kernel, iterations=3)

# Subtract sure foreground from sure
# background to get unknown region:
unknown = cv2.subtract(sure_bg, sure_fg)

# Label markers:
_, markers = cv2.connectedComponents(sure_fg)

# Add one to all labels so that
# sure background is not 0, but 1:
markers = markers + 1

# Mark the unknown regions with zero:
markers[unknown == 255] = 0

# Convert the original grayscale
# image to a BGR color image:
colored_image = cv2.cvtColor(image, cv2.COLOR_GRAY2BGR)

# Apply the watershed algorithm:
markers = cv2.watershed(colored_image, markers)

# Mark watershed boundaries in red:
colored_image[markers == -1] = [255, 0, 0]

# Create a color image
# to display the results:
segmented_image = np.zeros_like(colored_image)

# Map each marker to
# a random color:
colors = []
for marker in range(2, np.max(markers) + 1):
    colors.append(np.random.randint(0, 255, size=3))
for marker, color in zip(range(2, np.max(markers) + 1), colors):
    segmented_image[markers == marker] = color

# Combine the segmented
# image with the original
# image:
final_image = cv2.addWeighted(colored_image, 0.4, segmented_image, 0.6, 0)

# Display the original and
# watershed segmented images:
plt.figure(figsize=(12, 6))

plt.subplot(1, 2, 1)
plt.imshow(cv2.cvtColor(colored_image, cv2.COLOR_BGR2RGB))
plt.title('Original Image with Boundaries')
plt.axis('on')

plt.subplot(1, 2, 2)
plt.imshow(cv2.cvtColor(final_image, cv2.COLOR_BGR2RGB))
plt.title('Watershed Segmentation')
plt.axis('on')

plt.show()
```

Python implementation 17: Watershed algorithm.

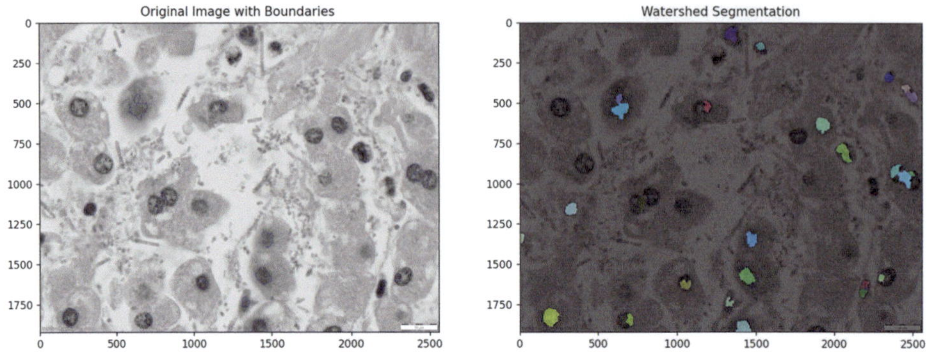

Fig. 5.4 The watershed algorithm applied to images showing liver parenchyma of a pig infected with *Clostridium* spp. Images were taken at 100× magnification, with a scale bar representing 10 micrometers. The left panel shows the original histological image, highlighting the hepatic parenchyma and areas affected by the bacterial infection. The right panel displays the image after processing with the watershed algorithm, which is used to segment and delineate cellular and structural boundaries, enhancing the visualization of infected regions and the organization of hepatic tissue

To separate overlapping objects, the *watershed* algorithm is employed. This method treats the grayscale image as a *topographic surface*, where bright pixels are considered peaks and dark pixels valleys. The algorithm simulates the flooding of these valleys, with boundaries being established where different floods meet, effectively delineating the objects. The code marks these boundaries on the original image and assigns random colors to each segmented region, making it easier to distinguish between different structures within the tissue (Fig. 5.4). This technique is invaluable in the context of veterinary pathology, for isolating and analyzing individual cells or structures within densely packed tissue samples, where traditional segmentation methods might struggle. For example, in tumor analysis, the watershed algorithm can help separate overlapping cells, enabling precise measurement of tumor size, cell density, and the extent of infiltration into surrounding tissues. This may be crucial for accurate staging of cancers and for guiding treatment decisions.

Moreover, this approach is highly beneficial in research settings, where large-scale analysis of tissue morphology is required. The ability to automatically segment and quantify different regions within a sample accelerates the analysis process and enhances the reproducibility of results, which is critical in studies investigating the effects of diseases or treatments on tissue structure. For example, my colleagues and I used a very similar methodology in 2014 to determine the structure of the healthy human pancreas, where colors determined peaks (high islet density) and valleys (low islet density) [118].

5.3.3 Preprocessing Steps for Segmentation

The code provided earlier (Python Implementation 5.4) implements an advanced image segmentation technique using the *watershed* algorithm, which is particularly useful in separating overlapping structures within histological images. However, to achieve more accurate segmentation results, especially in complex tissue samples like those encountered in veterinary pathology, several preprocessing steps can be enhanced. First, contrast enhancement is crucial because histological images often exhibit subtle differences in intensity that can make it challenging for the watershed algorithm to distinguish between different regions. The contrast between different tissue structures can be improved by applying *contrast-limited adaptive histogram equalization* (CLAHE), making them more distinguishable in the segmentation process [119]. Noise reduction is another vital step. While Gaussian blurring is effective at reducing noise, more sophisticated techniques like *nonlocal means denoising* or *median blurring* might be more appropriate for preserving the edges of important structures while removing background noise [112]. This ensures that the segmentation focuses on the relevant regions of the tissue sample. The accuracy of the *watershed* algorithm largely depends on the gradient map used to define the boundaries between different regions. The gradient magnitude can be calculated by applying a *Sobel* filter (discussed earlier), which enhances the edges within the image, making the boundaries clearer for the algorithm to detect [120]. Moreover, the markers used to indicate foreground and background regions play a critical role in the effectiveness of the watershed algorithm. Refining these markers with additional morphological operations like erosion and dilation can better define the areas that are to be segmented. This refinement helps the algorithm accurately separate the foreground (e.g., cells or tissue regions) from the background. Edge detection can also be combined with the *watershed* algorithm to further improve the segmentation. Of note is that the algorithm can more precisely delineate the boundaries between different regions by detecting edges within the enhanced image and using these edges to define the markers [120]. In veterinary pathology, accurate segmentation of histological images is crucial for tasks such as measuring tumor size, analyzing tissue architecture, and identifying pathological changes in tissue samples. These enhancements to the preprocessing steps can lead to more reliable segmentation, allowing pathologists to draw more accurate conclusions from the analysis. This is particularly important in research settings, where the ability to automatically and accurately segment large numbers of tissue samples can significantly speed up the analysis process and improve the reproducibility of results [118].

5.4 Morphological Operations and Blob Detection

Morphological operations are fundamental image processing techniques that focus on the structure or morphology of features within an image. These operations include processes such as *dilation*, *erosion*, *opening*, and *closing* [121]. *Dilation* expands the boundaries of objects in an image, making features larger and more prominent. *Erosion*, on the other hand, shrinks these boundaries, effectively reducing the size of features. Opening and closing are combinations of *erosion* and *dilation*; opening removes small objects from an image, while closing fills in small holes within objects. These operations are particularly useful in preparing images for further analysis, such as segmentation or feature extraction, by refining the shapes and structures within the image. For example, they can help eliminate noise or small irrelevant details, making it easier to identify and analyze the significant regions of interest. *Blob* detection is another crucial technique in image processing, designed to identify and analyze regions in an image that differ in properties such as intensity or color compared to surrounding regions. These regions, known as blobs, could represent objects, cells, or other relevant structures in a histological image. Techniques like *Laplacian of Gaussian* (LoG), *difference of Gaussians* (DoG), and *determinant of Hessian* are commonly used for blob detection.

5.4.1 Enhanced Edges and Dilation

Let us take an example that performs a sequence of image processing steps on a histological image, which are useful for analyzing tissue samples in veterinary medicine. The code from Python Implementation 5.5 begins by loading an image of a histological slide from the specified file path using *OpenCV*. It then converts the loaded image to grayscale. The conversion of an image to grayscale simplifies further processing by reducing the image to a single channel, where intensity variations are represented by shades of gray. This is a common preprocessing step when the color information is not critical for the analysis.

Enhanced edges and dilation

```
import cv2
import matplotlib.pyplot as plt

# Load the image by
# using the specified path.

image_path = r'C:\Users\Elitebook\Desktop\his_img.jpg'
image = cv2.imread(image_path)

# Convert to grayscale.
gray_image = cv2.cvtColor(image, cv2.COLOR_BGR2GRAY)

# Optional: Enhance contrast
# using histogram equalization.

equalized_image = cv2.equalizeHist(gray_image)

# Apply Gaussian Blur.
blurred_image = cv2.GaussianBlur(equalized_image, (5, 5), 0)

# Apply Canny Edge Detection
# with adjusted (higher) thresholds.
# Higher thresholds for more prominent edges.

edges = cv2.Canny(blurred_image, 100, 200)

# Display the results.
plt.figure(figsize=(15, 5))

plt.subplot(1, 3, 1)
plt.imshow(gray_image, cmap='gray')
plt.title('Grayscale Image')
plt.axis('on')

plt.subplot(1, 3, 2)
plt.imshow(blurred_image, cmap='gray')
plt.title('Blurred Image')
plt.axis('on')

plt.subplot(1, 3, 3)
plt.imshow(edges, cmap='gray')
plt.title('Edge Detection')
plt.axis('on')

plt.show()
```

Python implementation 18: Enhanced edges and dilation.

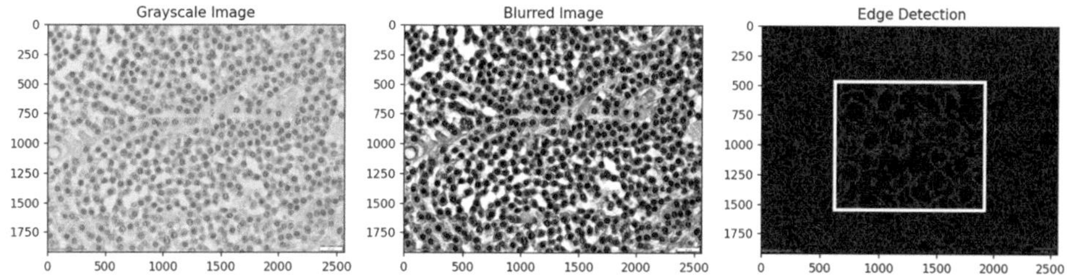

Fig. 5.5 Enhanced edges and dilation applied to an image showing a mast cell tumor (mastocytoma) in the skin of a cat. Images were captured at 40× magnification, with a scale bar representing 20 micrometers. The left panel shows the original histological image, highlighting the dense population of mast cells characteristic of the tumor. The middle panel displays the processed and blurred image, where Gaussian blur has been applied to reduce noise and smooth the image while maintaining the general structure of the tumor cells. The right panel presents the processed image using Canny edge detection, which emphasizes the boundaries and edges of the mast cells and surrounding tissue structures, aiding in the delineation of cellular and architectural features within the tumor. A zoom-in windows is included in the right image to provide reference points on the printed page, allowing for detailed examination

Next, the code optionally enhances the contrast of the grayscale image using histogram equalization. This step helps to improve the visibility of structures within the tissue by spreading out the intensity values, making it easier to distinguish different features that may be important for diagnostic purposes. Following contrast enhancement, the code applies a *Gaussian blur* to the image (Python Implementation 5.5). Blurring smoothens the image by reducing noise and small, irrelevant details, which can interfere with subsequent analysis steps (Fig. 5.5). In this case, the blurring helps to prepare the image for edge detection by ensuring that the detected edges represent significant transitions in intensity rather than noise [120].

The key processing step is the application of *Canny edge* detection [120]. This algorithm identifies areas of the image where there is a sharp change in intensity, which typically corresponds to the boundaries of structures within the tissue. The *Canny* method is particularly powerful because it includes both a gradient computation and non-maximum suppression, which helps to highlight only the most prominent edges. The code uses relatively high threshold values to focus on more significant edges, which are likely to correspond to important morphological features such as cell boundaries or tissue interfaces. Next, the code visualizes the original grayscale image, the blurred image, and the detected edges side by side using, as many times before, the *Matplotlib* library (Fig. 5.5). This allows for an easy comparison of how each processing step modifies the image and highlights different aspects of the tissue structure. This type of image processing is invaluable. It enables veterinarians and pathologists to enhance and isolate key features within tissue samples, facilitating the identification of pathological changes such as abnormal cell structures, tissue organization, or foreign elements.

5.4.2 Edge Detection Using Sobel Operator

The *Sobel* edge detection technique stands as a unique segmentation method [122, 123]. The code from Python Implementation 5.6 applies the *Sobel* edge detection technique to a histological image, similar to the previous code that utilized *Canny* edge detection. In this implementation, the *Sobel* operator is used to detect edges by calculating the gradient of image intensity in both the horizontal (*X*) and vertical (*Y*) directions (Python Implementation 5.6). The *Sobel* operator highlights areas of rapid intensity change, which typically correspond to edges within the image [124]. The *Sobel* edge detection is performed separately in the *X* and *Y* directions using the *cv2.Sobel* function with a kernel size of 5 (Fig. 5.6). The results are stored in *sobelx* and *sobely*, which represent the gradients in the horizontal and vertical directions, respectively. To combine these two directional gradients into a single representation of edge intensity, the code calculates the magnitude of the gradient using *cv2.magnitude*. This combined *Sobel* output is then normalized to a 0–255 range to ensure that the image can be properly visualized as an 8-bit image, making the edges more discernible.

Edge detection using sobel operator

```
import cv2
import numpy as np
import matplotlib.pyplot as plt

# Load the image:
image_path = r'C:\Users\Elitebook\Desktop\his_img.jpg'
image = cv2.imread(image_path)

# Convert to grayscale.
gray_image = cv2.cvtColor(image, cv2.COLOR_BGR2GRAY)

# Apply Sobel Edge Detection.
sobelx = cv2.Sobel(gray_image, cv2.CV_64F, 1, 0, ksize=5)  # X-direction.
sobely = cv2.Sobel(gray_image, cv2.CV_64F, 0, 1, ksize=5)  # Y-direction.
sobel_combined = cv2.magnitude(sobelx, sobely)

# Normalize the Sobel result to 0-255 range.
sobel_combined = np.uint8(255 * sobel_combined / np.max(sobel_combined))

# Show results:
plt.figure(figsize=(18, 6))

plt.subplot(1, 3, 1)
plt.imshow(sobelx, cmap='gray')
plt.title('Sobel X')
plt.axis('on')

plt.subplot(1, 3, 2)
plt.imshow(sobely, cmap='gray')
plt.title('Sobel Y')
plt.axis('on')

plt.subplot(1, 3, 3)
plt.imshow(sobel_combined, cmap='gray')
plt.title('Combined Sobel Edge Detection')
plt.axis('on')

plt.tight_layout()
plt.show()
```

Python implementation 19: Edge detection using sobel operator.

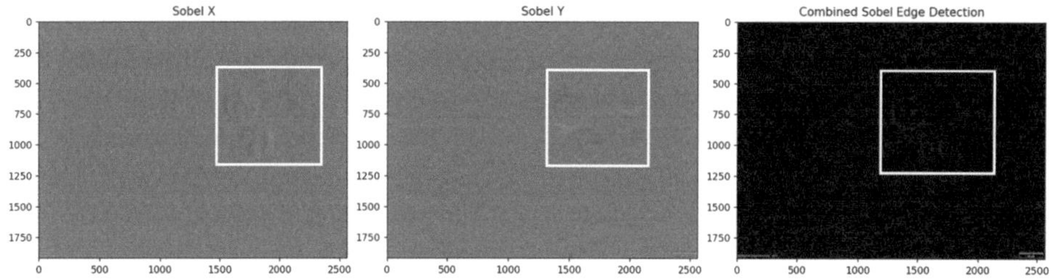

Fig. 5.6 Edge detection using Sobel operator applied to an image showing a mast cell tumor (mastocytoma) in the skin of a cat. Images were captured at 40× magnification, with a scale bar representing 20 micrometers. The left panel displays the Sobel X image, highlighting the horizontal edges in the histological section. The middle panel shows the Sobel Y image, emphasizing the vertical edges. The right panel presents the processed image using combined Sobel edge detection, which integrates both horizontal and vertical edge information to provide a comprehensive view of the cellular and structural boundaries within the tumor. Zoom-in windows are included in each image to provide reference points on the printed page, allowing for detailed examination of specific areas

In comparison to *Canny* edge detection, the *Sobel* operator provides a more direct calculation of the gradient and can capture finer details of the gradient changes in both directions, although it may not always provide the same level of edge refinement and noise suppression that *Canny* offers. The output is three images: the horizontal gradient (*Sobel X*), the vertical gradient (*Sobel Y*), and the combined gradient magnitude (combined *Sobel edge* detection).

This approach is useful in veterinary medicine as it helps in the analysis of histological images by highlighting structures within tissue samples, such as cellular boundaries and tissue morphology, much like the other methods described earlier. However, edge detection techniques like *Sobel* can aid in the identification and quantification of cellular features, which are critical for diagnosing various conditions, assessing tissue integrity, and understanding pathological changes. Sobel edge detection can be particularly useful in identifying **subtle transitions** in tissue that may not be as apparent with other methods, thereby providing a complementary tool to other image processing techniques like *Canny* or morphological operations. This enhances the ability of veterinary professionals to perform more accurate and detailed analyses of tissue samples, supporting diagnostics and research in veterinary pathology.

5.5 Texture, Shape, and Color Feature Extraction

Texture, shape, and color feature extraction are also critical steps in the analysis of histological images, particularly in the context of veterinary pathology where these features can provide essential diagnostic information. Texture features describe the surface characteristics and patterns within an image, such as the regularity, coarseness, or smoothness of tissue structures. Techniques like the *gray-level co-occurrence matrix* (GLCM), *local binary patterns* (LBP), and *Gabor* filters are commonly used to quantify texture [125–127]. These methods help to capture subtle variations in pixel intensities that reflect underlying tissue properties, which can be indicative of specific pathological conditions or tissue types. *Shape* feature extraction focuses on identification and quantification of the geometric properties of objects within an image [128]. This methodology includes measurements such as area, perimeter, circularity, aspect ratio, and the identification of specific contours or boundary features using methods like contour detection and *Fourier* descriptors. Shape features are particularly useful for distinguishing between different types of cells, nuclei, or tissue structures and for identifying abnormalities such as irregular growth patterns or cellular morphologies that deviate from normal. *Color* feature extraction involves analyzing the distribution and relationships of colors within an image, which is especially important in histological staining techniques that use different dyes to highlight various tissue components. The extraction of color histograms and the use of color deconvolution, or the extraction of spatial distribution of colors, can lead to the possibility to differentiate between cell types, assess the degree of staining, and evaluate the presence of pathological markers. Moreover, the color features can provide insights into the **biochemical** composition of tissues and are often used in conjunction with texture and shape features to enhance the overall analysis. Therefore, a working example is shown in Python Implementation 5.7.

5.5.1 Multiple Parameters Used as Tissue Signatures

The main code snippet from Python Implementation 5.7 performs feature extraction from a histological image using three primary methods: *texture analysis*, *shape analysis*, and *color analysis*. These methods are critical in histological image analysis for veterinary pathology, where detailed quantification of tissue characteristics is necessary for diagnosing diseases and understanding tissue morphology. Of note is the preparation before Python Implementation 5.7 is run. That is, if the code is directly run for the first time, a package installation is needed first, such as PIP:

```
pip install opencv-python-headless matplotlib scikit-image numpy
```

Texture Feature Extraction uses the *gray-level co-occurrence matrix* (GLCM) to quantify the texture of the image. The GLCM calculates how often pairs of pixel intensities occur in a specified spatial relationship, capturing spatial variations in pixel values. The texture features extracted include *contrast*, *dissimilarity*, *homogeneity*, *energy*, and

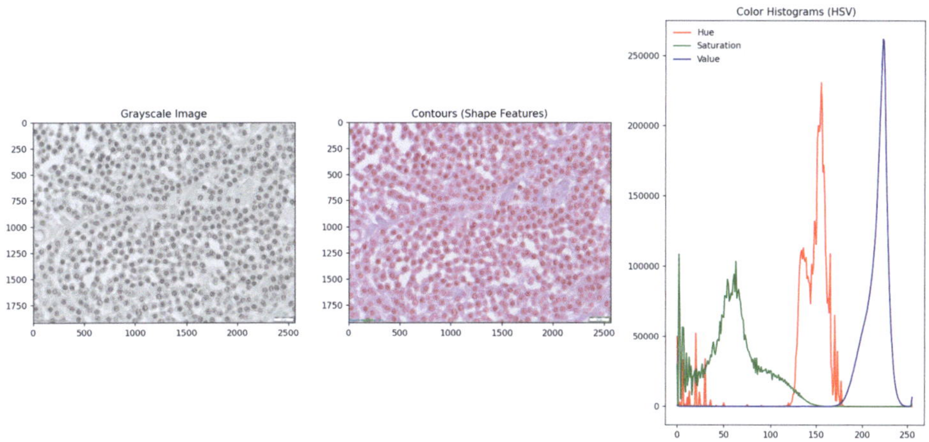

Fig. 5.7 Texture, shape, and color feature extraction applied to an image showing a mast cell tumor (mastocytoma) in the skin of a cat. Images were captured at 40× magnification, with a scale bar representing 20 micrometers. The left panel displays the grayscale image of the tumor, which highlights the cellular structures and texture details crucial for identifying mast cells and their morphological variations. The middle panel shows the contour image representing shape features extracted from the grayscale image, emphasizing the boundaries, circularity, and distribution of the cells, which are important for differentiating between normal and abnormal mast cell proliferation. The right panel presents a color histogram chart (HSV—hue, saturation, value), showing the distribution of color intensities across the tumor section, which can help in understanding the staining patterns and identifying different cellular components or abnormalities in the tissue. Together, these panels provide a comprehensive overview of texture, shape, and color characteristics for better diagnostic accuracy in mast cell tumors

correlation. In this analysis, the contrast value of 40.72 indicates significant variations in intensity, suggesting heterogeneous tissue structures (please see the output of Python Implementation 5.7). The dissimilarity value of 3.24 reflects a moderate level of variation between pixel pairs, while the homogeneity value of 0.31 indicates less uniformity, typical in complex tissue structures. The energy value of 0.033 suggests low textural uniformity, and the high correlation value of 0.97 implies that pixel intensities are strongly correlated, pointing towards predictable patterns within the image. These parameters are crucial for identifying specific histological patterns, such as cellular organization and texture irregularities, which can be indicative of pathological conditions like fibrosis or neoplasia in veterinary tissues. *Shape* Feature Extraction is performed by detecting contours within the image.

Texture, shape, and color feature extraction

```
import cv2
import numpy as np
from skimage.feature import graycomatrix, graycoprops, local_binary_pattern
from skimage.measure import regionprops, label
import matplotlib.pyplot as plt

# Load the image
image_path = r'C:\Users\Elitebook\Desktop\his_img.jpg'  # Your local path
image = cv2.imread(image_path)

# Convert to grayscale
gray_image = cv2.cvtColor(image, cv2.COLOR_BGR2GRAY)

# Texture Feature Extraction using GLCM (Gray Level Co-occurrence Matrix)
# Calculate GLCM at multiple angles
glcm = graycomatrix(gray_image, distances=[1], angles=[0, np.pi/4, np.pi/2, 3*np.pi/4],
symmetric=True, normed=True)

# Extract texture features from GLCM
contrast = graycoprops(glcm, 'contrast')
dissimilarity = graycoprops(glcm, 'dissimilarity')
homogeneity = graycoprops(glcm, 'homogeneity')
energy = graycoprops(glcm, 'energy')
correlation = graycoprops(glcm, 'correlation')

print("Texture Features (GLCM):")
print(f"Contrast: {contrast.mean()}")
print(f"Dissimilarity: {dissimilarity.mean()}")
print(f"Homogeneity: {homogeneity.mean()}")
print(f"Energy: {energy.mean()}")
print(f"Correlation: {correlation.mean()}")

# Shape Feature Extraction using Contour Detection
# Apply binary threshold
_, binary_image = cv2.threshold(gray_image, 127, 255, cv2.THRESH_BINARY)

# Find contours
contours, _ = cv2.findContours(binary_image, cv2.RETR_EXTERNAL, cv2.CHAIN_APPROX_SIMPLE)

# Draw contours and calculate shape features
shape_features = []
for contour in contours:
    area = cv2.contourArea(contour)
    perimeter = cv2.arcLength(contour, True)
    circularity = (4 * np.pi * area) / (perimeter ** 2) if perimeter != 0 else 0
    shape_features.append((area, perimeter, circularity))

print("Shape Features:")
print(f"Number of contours: {len(contours)}")
for i, (area, perimeter, circularity) in enumerate(shape_features[:5]):  # Display first 5 for
brevity
    print(f"Contour {i+1} - Area: {area}, Perimeter: {perimeter}, Circularity: {circularity}")

# Color Feature Extraction using Color Histograms
# Convert to HSV and extract histograms for each channel
hsv_image = cv2.cvtColor(image, cv2.COLOR_BGR2HSV)
h_hist = cv2.calcHist([hsv_image], [0], None, [256], [0, 256])
s_hist = cv2.calcHist([hsv_image], [1], None, [256], [0, 256])
v_hist = cv2.calcHist([hsv_image], [2], None, [256], [0, 256])

print("Color Features (HSV Histograms):")
print(f"Hue Mean: {np.mean(h_hist)}")
print(f"Saturation Mean: {np.mean(s_hist)}")
print(f"Value Mean: {np.mean(v_hist)}")

# Display images with annotations
plt.figure(figsize=(15, 5))

plt.subplot(1, 3, 1)
plt.imshow(gray_image, cmap='gray')
plt.title('Grayscale Image')
plt.axis('on')

plt.subplot(1, 3, 2)
plt.imshow(cv2.drawContours(image.copy(), contours, -1, (0, 255, 0), 1))
plt.title('Contours (Shape Features)')
plt.axis('on')

plt.subplot(1, 3, 3)
plt.plot(h_hist, color='r', label='Hue')
plt.plot(s_hist, color='g', label='Saturation')
plt.plot(v_hist, color='b', label='Value')
plt.title('Color Histograms (HSV)')
plt.legend()
plt.axis('on')

plt.tight_layout()
plt.show()
```

Python implementation 20: Texture, shape, and color feature extraction.

The contours represent the boundaries of various objects or structures, such as cells or nuclei. Here, the analysis yielded 26 contours, which provide a measure of the number of distinct objects detected in the image. The extracted shape features include area, perimeter, and circularity for each contour. Of note is that the contours detected may not be the expected contour of the cells from the image, but rather higher-level organizations of cell structures. For instance, contour 1 has an area of 1.0 and a perimeter of 4.0, with a circularity of 0.785, indicating it is close to circular. In contrast, contour 5 has a minimal area of 0.5, a perimeter of 15.41, and a very low circularity of 0.026, suggesting a highly irregular shape. These shape features are valuable for identifying and characterizing the **morphological aspects of cell structures**, which are essential in assessing cellular health, detecting abnormalities, and classifying cell types. In veterinary pathology, these shape metrics help in identifying abnormal cell morphologies that could be indicative of dysplasia, malignancy, or inflammatory conditions.

Output

```
Texture Features (GLCM):
Contrast:       40.72082225716996
Dissimilarity: 3.236542709315649
Homogeneity:    0.309004258648409
Energy:         0.033931610158769
Correlation:    0.974862743544512

Shape Features:
Number of contours: 26
Contour 1 - Area: 1.0, Perimeter: 4.0, Circularity: 0.7853981633974483
Contour 2 - Area: 3.0, Perimeter: 8.0, Circularity: 0.5890486225480862
Contour 3 - Area: 5.5, Perimeter: 11.071067690849304, Circularity: 0.5638888941900405
Contour 4 - Area: 0.0, Perimeter: 0.0, Circularity: 0
Contour 5 - Area: 0.5, Perimeter: 15.41421353816986, Circularity: 0.0264446075409239

Color Features (HSV Histograms):
Hue Mean:          19200.0
Saturation Mean: 19200.0
Value Mean:        19200.0
```

Color feature extraction involves converting the image into the HSV (*hue*, *saturation*, *value*) color space and analyzing the histograms of each channel (Fig. 5.7). The mean values of hue, saturation, and value are all 19200.0, suggesting a uniform distribution across these channels. The hue histogram reflects the predominant colors in the image, which are directly related to the staining characteristics of the tissue sample. Stains such as hematoxylin and eosin are commonly used in histology to highlight different cellular components, with hematoxylin staining the nuclei blue and eosin staining the cytoplasm and extracellular matrix pink. Saturation measures the intensity of the color, while value reflects the brightness [119]. Variations in these color parameters can reveal differences in tissue composition and the presence of pathological changes. In veterinary pathology, accurate color feature extraction can assist in differentiating between normal and abnormal tissue regions, aiding in the diagnosis of conditions such as neoplasia, inflammation, and necrosis.

Nevertheless, the integration of *texture*, *shape*, and *color* feature extraction provides a comprehensive approach to analyzing histological images in veterinary medicine [129]. These features enable the quantitative assessment of tissue characteristics, facilitating objective evaluation and enhancing the accuracy of diagnostic processes [130]. Veterinary pathologists can efficiently process large volumes of data by automating the extraction and analysis of these features, leading to more reliable diagnoses and better-informed treatment decisions for animal patients. The scientific rigor of these methods supports their use in both clinical and research settings, advancing the field of veterinary pathology through improved image analysis techniques.

Quantitative Image Analysis in Veterinary Medicine

6

Abstract

Quantitative image analysis in veterinary medicine has revolutionized the diagnosis and prognosis of various animal diseases by providing objective and reproducible data from microscopic images. This chapter explores novel practical examples in Python, for some transformative computational techniques on traditional pathology, specifically through automated analysis of biological tissues and cells. Quantitative methods offer significant advantages over subjective visual interpretation, including the ability to accurately measure cell density, morphology, and spatial distribution. These parameters are crucial for diagnosing conditions such as cancers and inflammatory disorders, where changes at the cellular level are indicative of disease. Advanced segmentation algorithms, texture analysis, and feature extraction further enable detailed tissue characterization at both microscopic and macroscopic levels. The integration of deep learning and artificial intelligence into image analysis has expanded diagnostic capabilities, allowing for the detection of complex patterns associated with disease progression. Automated cell counting, morphological measurements, and staining intensity quantification significantly enhance diagnostic accuracy while reducing human error. The chapter also discusses the applications of these techniques in research and education, where they contribute to a better understanding of disease mechanisms and training of veterinary professionals.

Keywords

Quantitative analysis · Cell counting · Morphological parameters · Staining intensity · Statistical analysis

E. Gagniuc, *Computational Analysis of Microscopy Images*,
https://doi.org/10.1007/978-3-031-86270-0_6

6.1 Introduction

Quantitative image analysis in veterinary medicine represents a transformative approach to understanding and diagnosing a wide array of animal diseases. Unlike traditional methods that rely heavily on qualitative assessments and subjective interpretations by pathologists, quantitative image analysis utilizes computational techniques to objectively measure and analyze biological tissues and cells from microscopic images. This approach leverages advanced technologies such as image processing, machine learning, and artificial intelligence to extract meaningful data from digital images, enabling more precise and reproducible evaluations. The integration of quantitative image analysis into veterinary medicine provides numerous benefits. For instance, it allows for the detailed examination of histological samples, including the quantification of cell density, morphology, and spatial distribution. These parameters are crucial for diagnosing conditions such as cancers, infectious diseases, and inflammatory disorders, where changes at the cellular level can provide critical insights into the nature and progression of the disease. Automated systems that detect and quantify such changes can significantly reduce human error, increase diagnostic accuracy, and enhance workflow efficiency in clinical and research settings. Furthermore, quantitative image analysis is not confined to simple measurements. It encompasses a range of sophisticated methods, including segmentation, texture analysis, and feature extraction, that enable the detailed characterization of tissues at both the microscopic and macroscopic levels. Advanced segmentation algorithms can precisely delineate cellular boundaries and isolate specific tissue types, while texture analysis can assess the heterogeneity within a tissue sample, providing a more comprehensive understanding of tissue architecture. These techniques are particularly valuable in veterinary pathology, where tissue samples can be highly heterogeneous and traditional methods may fall short in providing detailed and quantifiable information. The advent of *deep learning* and *artificial intelligence* has further expanded the capabilities of quantitative image analysis in veterinary medicine. *Convolutional neural networks* (CNNs) and other *deep learning* models can be trained to recognize complex patterns and anomalies in histological images. Such models can be continually refined and validated against large datasets, allowing them to learn and improve over time. This iterative process leads to more robust and reliable diagnostic tools, which are becoming indispensable in modern veterinary practices. The application of quantitative image analysis is also crucial in veterinary research, where it plays a pivotal role in drug development, toxicology studies, and understanding disease mechanisms. Objective and reproducible data enable researchers to compare results across different studies, track changes over time, and make data-driven decisions that enhance the validity and impact of their findings. In educational settings, these techniques offer a valuable resource for training veterinary students, helping them develop a deeper understanding of pathology and diagnostic processes.

6.1.1 The Importance of Quantitative Image Analysis

Traditional microscopic analysis often relies on the subjective interpretation of visual information by a trained pathologist. While this approach has been the cornerstone of veterinary diagnostics for decades, it is inherently limited by variability in interpretation, potential human error, and difficulty in quantifying subtle differences across samples. Quantitative image analysis addresses these challenges by converting visual data into measurable parameters, thereby providing a more objective and standardized assessment. In veterinary medicine, where accurate diagnosis and prognosis are critical for animal health, the application of quantitative image analysis enables the precise counting of cells, measurement of morphological features, and analysis of staining intensities.

6.1.2 Applications in Veterinary Pathology

Quantitative image analysis is utilized across a broad spectrum of veterinary pathology applications [86]. It is essential for tasks such as cell counting, which is crucial in diagnosing conditions like anemia or leukocytosis. Similarly, measuring the size, shape, and distribution of cells or tissues can reveal insights into pathological changes, such as tumor grading or the identification of inflammatory responses [131, 132]. Another vital application is the quantification of staining intensity in histological samples, which helps assess the presence and extent of specific biomarkers [133]. This is particularly useful in immunohistochemistry, where the precise measurement of protein expression levels can inform diagnostic and therapeutic decisions. Beyond cell and tissue assessments, quantitative image analysis plays a role in larger-scale studies, including the statistical analysis of population-level data.

6.1.3 Challenges and Future Directions

Despite its many advantages, quantitative image analysis in veterinary medicine faces several challenges. These include the need for high-quality, standardized images, development of robust algorithms that can handle the complexity of biological tissues, and integration of quantitative data with other clinical information. Additionally, the adoption of these techniques in routine veterinary practice requires training and a shift in how pathologists and clinicians approach diagnostics. Looking forward, the future of quantitative image analysis in veterinary medicine is promising. Emerging technologies, such as three-dimensional (3D) imaging and advanced AI-driven analytics, are expected to further enhance the precision and applicability of these methods. As these tools become more accessible and integrated into clinical workflows, they will likely play an increasingly central role in veterinary diagnostics, research, and personalized medicine for animals.

6.2 Counting Cells and Structures

Counting cells and structures is a fundamental aspect of quantitative image analysis in veterinary medicine, providing critical information for diagnosing diseases, evaluating tissue health, and conducting research [134, 135]. This process involves the automated or semiautomated identification and enumeration of cells, nuclei, and other histological features within a digital image, which is crucial for assessing cellular density, proliferation, and architectural organization. Accurate cell counting can reveal changes in tissue composition and organization that may indicate pathological conditions such as cancer, inflammation, or infection [136]. Automated cell counting techniques employ image processing algorithms that can detect and differentiate cells based on their unique visual characteristics, such as shape, size, and staining properties. The first step often involves converting the image to grayscale and applying thresholding methods like *Otsu* thresholding to distinguish foreground structures from the background [89]. Morphological operations, such as erosion, dilation, and closing, are then used to refine the segmented areas, ensuring that individual cells or structures are well defined and separated from each other [137, 138]. Advanced segmentation techniques, such as watershed segmentation, region growing, and deep learning models like *U-Net*, can further improve the accuracy of cell counting by precisely delineating the boundaries of closely packed or overlapping cells. These methods leverage texture, intensity, and spatial information to accurately segment each cell or structure, even in challenging images with varying staining intensities and complex tissue architectures. Machine learning approaches can be trained to recognize and count different cell types or structures, enhancing the flexibility and adaptability of the analysis process. Quantifying cellular and structural features goes beyond mere counting. It can include calculating additional parameters like cell size, shape, circularity, and distance between neighboring cells. These measurements provide a more comprehensive understanding of the microenvironment of the tissue, which is essential for distinguishing between benign and malignant growths, identifying hyperplasia or hypertrophy, and evaluating the effects of treatments in both clinical and research settings. Counting structures and cells is not limited to histopathology. It also extends to other applications, such as assessing immune cell infiltration in immunohistochemistry, evaluating synaptic densities in neurological studies, and quantifying bacteria in microbiology. Without question, the ability to consistently and accurately count cells and structures in diverse contexts makes quantitative image analysis a powerful tool in veterinary diagnostics and research.

6.2.1 Automated Cell Nuclei Counting in Histology

Here, a new example demonstrates an effective workflow for segmenting and counting cell nuclei in histological images using thresholding and morphological operations (Python Implementation 6.1). The method provides a reliable and efficient means of quantifying cellular structures by using the unique properties of nuclei, such as their heterochromatin content and distinct staining patterns (which is of great value in veterinary and medical pathology). Thus, the code uses some image processing techniques described above, to automatically count cell nuclei in a histological image. This approach is highly valuable for quantitative analysis in histopathology, where cell counting is essential for diagnosing diseases, understanding tissue architecture, and conducting various biomedical research studies. Python Implementation 6.1 begins by loading a histological image and converting it to grayscale. This step is crucial because grayscale images simplify the data by reducing it to a single intensity channel, making subsequent processing more straightforward and computationally efficient. The next step involves applying *Otsu* thresholding (discussed in the previous chapters), a popular technique that automatically determines an optimal threshold value to separate the foreground (nuclei) from the background based on the grayscale histogram of the image. The *Otsu* method isolates the nuclei as white regions (foreground) against a black background by converting the image to a binary mask (Fig. 6.1).

Automated cell nuclei counting in histology

```
import cv2
import numpy as np
import matplotlib.pyplot as plt
from skimage.io import imread
from skimage.filters import threshold_otsu
from skimage.morphology import remove_small_objects, remove_small_holes, closing, square, label

# Load the original histological image.
image_path = r'C:\Users\Elitebook\Desktop\his_img.jpg'
image = imread(image_path)

# Convert to grayscale:
gray_image = cv2.cvtColor(image, cv2.COLOR_RGB2GRAY)

# Apply Otsu thresholding:
otsu_thresh = threshold_otsu(gray_image)
mask_otsu = (gray_image < otsu_thresh).astype(np.uint8) * 255

# Refine the mask with
# morphological operations.

mask_otsu_refined = remove_small_objects(mask_otsu.astype(bool), min_size=64).astype(np.uint8) * 255
# Remove small objects
mask_otsu_refined = remove_small_holes(mask_otsu_refined.astype(bool),
area_threshold=64).astype(np.uint8) * 255  # Fill small holes
# Further closing to smooth the mask.
mask_otsu_refined = closing(mask_otsu_refined, square(3))

# Count the white "islands" (nuclei of cells)
# Use connected components to find and count
# distinct white regions.

num_labels, labels, stats, centroids = cv2.connectedComponentsWithStats(mask_otsu_refined,
connectivity=8)

# Subtract 1 from the count to
# ignore the background label
# (label 0).

cell_count = num_labels - 1

# Visualize the refined mask.
plt.figure(figsize=(6, 6))
plt.imshow(mask_otsu_refined, cmap='gray')
plt.title(f'Refined Otsu Thresholding Mask - Cell Count: {cell_count}')
plt.axis('on')
plt.show()

print(f"Number of white islands (nuclei of cells) in the image: {cell_count}")
```

Python implementation 21: Automated cell nuclei counting in histology.

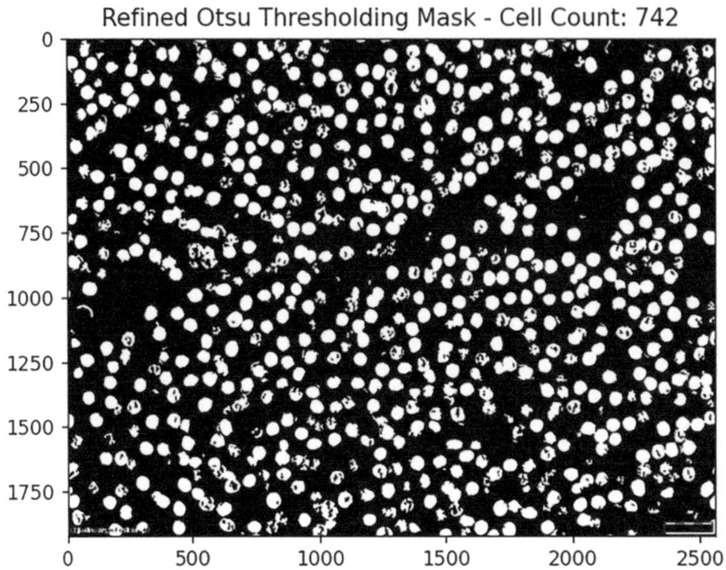

Fig. 6.1 Automated cell nuclei counting in histological images of a skin mast cell tumor (mastocytoma) in a cat. Images were acquired at 40× magnification, with a scale bar representing 20 μm. The image shows the binary mask obtained using Otsu thresholding, which is used for the nondiscriminatory quantification of all white islands (representing cell nuclei). This approach highlights the nuclei of all cells within the tumor tissue, enabling automated and unbiased counting of cell nuclei, which is crucial for assessing cell density and proliferation in pathological analysis of mast cell tumors

Following thresholding, the code refines the binary mask using morphological operations to enhance the accuracy of the segmentation. The *remove_small_objects* function eliminates noise or tiny artifacts that could be mistakenly identified as nuclei, ensuring that only significant structures are retained. The *remove_small_holes* operation fills in small gaps within the segmented regions, helping to create more contiguous representations of the nuclei. Finally, the closing operation, performed with a square structuring element, smoothens the boundaries of the segmented regions by bridging narrow breaks and closing small gaps, which further refines the appearance of the nuclei. The refined binary mask is then analyzed to count the distinct white regions, each representing a nucleus.

The *cv2.connectedComponentsWithStats* function is used to identify and label connected components in the binary image, which correspond to individual nuclei. This function returns several outputs: the total number of labels (including the background), a labeled image where each connected component is assigned a unique label, and statistics about each component, such as its area and centroid. The code excludes the background by subtracting 1 from the total count, providing an accurate count of the cell nuclei in the image (Python Implementation 6.1). Cell nuclei are ideal for counting cells because of their distinct visual characteristics, particularly their heterochromatin content, which makes them stand out in histological stains (Fig. 6.1). Heterochromatin, a tightly packed form of DNA, stains more intensely compared to the surrounding euchromatin, creating a high contrast between the nucleus and the cytoplasm or background tissue [139–141]. This contrast is essential for segmentation algorithms to accurately differentiate nuclei from other structures. Additionally, nuclei are typically round to oval in shape and have relatively consistent sizes within a given cell type, which makes them well suited for automated detection and counting using morphological and image processing techniques. The ability to accurately count cell nuclei in histological images is crucial in various applications. In pathology, it may help in assessing cell density and proliferation rates, which are key indicators of disease states such as cancer, where abnormal cell growth and division are common. In research, quantifying cell numbers can aid in evaluating the effects of drugs or other interventions on tissue samples. Automated cell counting, as demonstrated in this code, may reduce the time and potential for human error compared to manual counting, offering a more standardized and reproducible approach to tissue analysis.

6.2.2 Automated Cell Counting with Size-Based Filtering

The next code builds upon the previous code by incorporating additional steps to refine the process of counting cell nuclei in histological images (Python Implementation 6.2). While both codes share the same foundational approach of image preprocessing, thresholding, and morphological operations, the new code introduces a more sophisticated filtering mechanism that enhances the accuracy and reliability of cell counting. Thus, the primary difference between this code and the previous one is the introduction of a filtering criterion based on the size of the segmented "islands" (i.e., nuclei). After performing the *Otsu* thresholding and refining the mask through morphological operations, this code calculates the average size of all detected white islands (i.e., nuclei) and sets a minimum size threshold for what constitutes a valid nucleus. Specifically, it filters out any island whose size is smaller than half of the average size of all islands (note that any rule can be imposed, however). This additional step ensures that only significant structures are counted, excluding small artifacts or noise that might have been mistakenly included in the previous approach.

Automated Cell Counting with Size-Based Filtering

```
import cv2
import numpy as np
import matplotlib.pyplot as plt
from skimage.io import imread
from skimage.filters import threshold_otsu
from skimage.morphology import remove_small_objects, remove_small_holes, closing, square

image_path = r'C:\Users\Elitebook\Desktop\his_img.jpg'
image = imread(image_path)
gray_image = cv2.cvtColor(image, cv2.COLOR_RGB2GRAY)

# Otsu thresholding:
otsu_thresh = threshold_otsu(gray_image)
mask_otsu = (gray_image < otsu_thresh).astype(np.uint8) * 255

# Refine the mask with morphological
# operations as before:

mask_otsu_refined = remove_small_objects(mask_otsu.astype(bool), min_size=64).astype(np.uint8) * 255
mask_otsu_refined = remove_small_holes(mask_otsu_refined.astype(bool),
area_threshold=64).astype(np.uint8) * 255
mask_otsu_refined = closing(mask_otsu_refined, square(3))

# Count the white "islands" (nuclei of cells):
num_labels, labels, stats, centroids = cv2.connectedComponentsWithStats(mask_otsu_refined,
connectivity=8)

# Calculate the average size of the
# islands (excluding the background).
# Area of each label, skip label 0
# (background):

island_sizes = stats[1:, cv2.CC_STAT_AREA]
average_island_size = np.mean(island_sizes)

# Define minimum island size as
# half the average island size.
min_island_size = average_island_size / 2

# Filter out islands smaller
# than the minimum size.
valid_islands = [size for size in island_sizes if size >= min_island_size]

# Count the valid islands:
valid_island_count = len(valid_islands)

print(f"Number of valid white islands (nuclei of cells) in the image: {valid_island_count}")
print(f"Average island size: {average_island_size:.2f} pixels")
print(f"Minimum island size considered (average size / 2): {min_island_size:.2f} pixels")

plt.figure(figsize=(6, 6))
plt.imshow(mask_otsu_refined, cmap='gray')
plt.title(f'Refined Otsu Thresholding Mask - Valid Cell Count: {valid_island_count}')
plt.axis('on')
plt.show()
```

Python implementation 22: Automated Cell Counting with Size-Based Filtering.

The introduction of this filtering step is particularly valuable for improving the robustness of the cell counting process. In histological images, various factors such as staining intensity, tissue preparation, and imaging conditions can lead to the presence of small, irrelevant regions that do not represent actual nuclei but might still be detected during the segmentation process. The code effectively removes such noise by using the average island size as a benchmark and filtering out regions that fall below a defined threshold resulting in a more accurate count of valid nuclei. This is especially important when analyzing samples with high background variability or noise levels, where the risk of false positives is higher. Moreover, the code provides additional quantitative information that is valuable for further analysis and understanding of the tissue sample. It calculates and displays the average size of the islands, providing insights into the typical size of nuclei in the given sample. Knowing the average size and the minimum size considered helps in setting appropriate parameters for further analysis or for comparing different tissue samples under varying conditions.

Output

```
Number of valid white islands (nuclei of cells) in the image: 551
Average island size: 1739 pixels
Minimum island size considered (average size / 2): 870 pixels
```

From a scientific perspective, this refinement in the image analysis pipeline is beneficial because it enhances the quality of the data obtained from histological images. Accurate and reliable quantification of cell nuclei is crucial in veterinary and medical pathology for diagnosing diseases, assessing tissue health, and conducting research studies. Python Implementation 6.2 provides a cleaner, more focused analysis of the tissue, by excluding irrelevant regions that could skew the data, thus ensuring that the results are more representative of the biological reality (Fig. 6.2).

Additionally, this approach can be easily adapted and customized to accommodate different types of tissues, staining techniques, or imaging modalities, making it highly

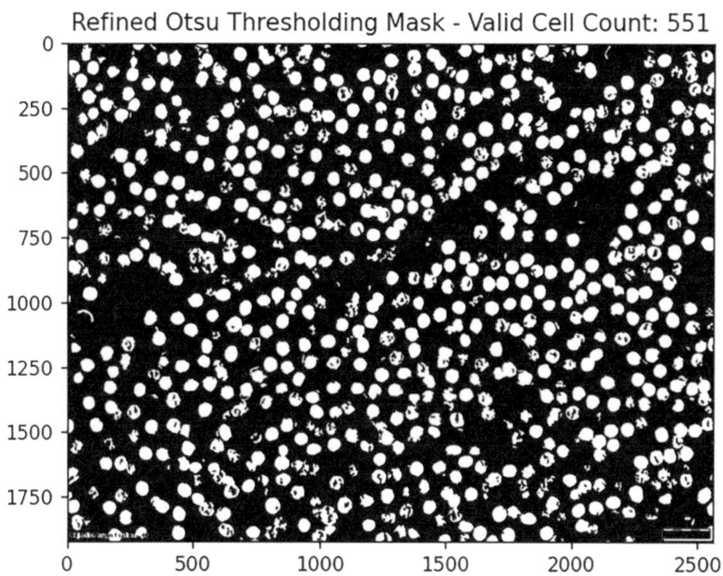

Fig. 6.2 Automated cell counting with size-based filtering in histological images of a mast cell tumor (mastocytoma) in cat skin. The image was captured at 40× magnification, with a scale bar representing 20 μm. The binary mask variant obtained through Otsu thresholding is shown, which is used to quantify all white islands (representing cell nuclei, totaling 551) larger than half the average size of an island. This refined rule aims to improve the accuracy of counting cell nuclei and, consequently, the number of cells by considering only the real nuclei, thereby avoiding false counts of residual artifacts

versatile and applicable to a wide range of histopathological studies. The refined counting mechanism, combined with the visualization of the refined mask, provides a comprehensive understanding of both the qualitative and quantitative aspects of the analyzed tissue, further supporting its utility in research and clinical practice. Nevertheless, the improvements introduced in this code make it a more reliable tool for automated cell counting, enhancing the accuracy of histopathological analyses and ensuring more robust and reproducible results in veterinary medicine and related fields.

6.3 Measuring Morphological Parameters

Measuring morphological parameters is a critical aspect of quantitative image analysis in veterinary medicine [135, 142]. It involves assessing the shape, size, structure, and spatial arrangement of cells and tissues within histological images [135]. These measurements provide valuable information for diagnosing various conditions, understanding disease progression, and evaluating tissue responses to treatments [135]. Veterinary pathologists can gain insights into the cellular architecture of normal and diseased tissues by analyzing morphological parameters, helping them identify abnormalities that may not be apparent through qualitative observation alone [135]. Several key morphological parameters are commonly measured in quantitative image analysis [135]. These include cell size, cell shape, nuclear-cytoplasmic ratio, perimeter, and circularity [135, 142]. Cell size and shape can indicate various conditions, such as hypertrophy, atrophy, or cellular atypia, often associated with inflammatory diseases, cancers, or other pathological changes [135, 143]. The nuclear-cytoplasmic ratio is particularly important in distinguishing between benign and malignant cells, as a higher ratio often suggests malignancy. Perimeter and circularity measurements can help assess cell integrity and regularity, providing additional clues to the health or pathology of a tissue sample [135]. Advanced computational tools and algorithms are employed to extract these morphological features from digital images [135]. Segmentation techniques, such as thresholding, edge detection, and deep learning–based models, play a vital role in accurately delineating cellular and subcellular structures. Once segmented, various mathematical models and shape descriptors are applied to quantify the morphological features. For example, shape descriptors like aspect ratio, roundness, and solidity can be calculated to evaluate cell and tissue shapes in more detail. These metrics are crucial for identifying subtle morphological differences that can inform disease diagnosis and treatment decisions. Texture analysis is another approach integrated into the measurement of morphological parameters. It involves quantifying the variation in pixel intensities within a region of interest to capture tissue heterogeneity. Textural features, such as contrast, entropy, and homogeneity, provide additional layers of information that complement traditional morphological measurements. The combination of texture analysis with morphological parameters enables a more comprehensive characterization of tissues, enhancing the diagnostic accuracy and specificity in veterinary pathology. Machine learning and artificial intelligence further enhance the measurement of morphological parameters by automating feature extraction and classification tasks. Algorithms such as random forests, *support vector machines*, and *neural networks* can be trained to recognize specific morphological patterns associated with various diseases. These models can predict outcomes, identify disease subtypes, and aid in developing personalized treatment plans for animals. They also help to streamline the diagnostic process, allowing for faster and more consistent assessments across different samples and cases. The measurement of morphological parameters is also essential for research purposes in veterinary science. It facilitates the study of cellular responses to different therapeutic interventions, the investigation of disease mechanisms, and the development of new diagnostic tools. By providing quantifiable data, these measurements enable researchers to perform statistical analyses that improve the reliability and reproducibility of their findings. Moreover, they can be used to establish reference databases for various species, tissues, and disease states, supporting comparative studies and advancing the understanding of veterinary pathology. The use of measurement of morphological parameters in routine veterinary practice offers significant advantages. It enhances the precision of diagnoses, supports evidence-based decision-making, and ultimately leads to better clinical outcomes. As the field of quantitative image analysis continues to evolve, the integration of advanced techniques for measuring morphological parameters will become increasingly important in advancing veterinary medicine, providing more accurate and comprehensive insights into animal health and disease.

6.3.1 Quantitative Analysis of Nuclei Size and Distribution

This following code builds on the previous example by adding an additional layer of analysis that measures the spatial distribution of cell nuclei within a histological image (Python Implementation 6.3). While it still follows the same steps of loading, thresholding, and refining a binary mask to identify "islands" (nuclei), it introduces the computation of distances between neighboring nuclei. This enhancement provides valuable spatial information that can be crucial in understanding tissue organization and diagnosing pathological conditions. Like the previous code, it begins by loading the histological image and converting it to grayscale, making the subsequent steps more straightforward. Then, *Otsu* thresholding is applied to create a binary mask where nuclei appear as white regions against a black background [89]. Morphological operations are then used to refine the mask, removing small artifacts and filling in gaps to ensure that only well-defined nuclei are retained. The code calculates the size of each detected nucleus and filters out those smaller than half of the average size, ensuring that only valid, significant structures are counted. This filtering is useful for excluding noise and irrelevant objects that could distort the analysis. Note that because the same input image was used, Python Implementation 6.3 produces the same figure as in the previous example (Fig. 6.2).

Quantitative Analysis of Nuclei Size and Distribution

```
import cv2
import numpy as np
import matplotlib.pyplot as plt
from skimage.io import imread
from skimage.filters import threshold_otsu
from skimage.morphology import remove_small_objects, remove_small_holes, closing, square
from scipy.spatial import distance

image_path = r'C:\Users\Elitebook\Desktop\his_img.jpg'
image = imread(image_path)
gray_image = cv2.cvtColor(image, cv2.COLOR_RGB2GRAY)

# Apply Otsu thresholding:
otsu_thresh = threshold_otsu(gray_image)
mask_otsu = (gray_image < otsu_thresh).astype(np.uint8) * 255

# Refine the mask with
# morphological operations:

mask_otsu_refined = remove_small_objects(mask_otsu.astype(bool), min_size=64).astype(np.uint8) * 255
# Remove small objects
mask_otsu_refined = remove_small_holes(mask_otsu_refined.astype(bool),
area_threshold=64).astype(np.uint8) * 255                # Fill small holes.
mask_otsu_refined = closing(mask_otsu_refined, square(3))  # smooth the mask.

# Count the white "islands" (nuclei of cells):
num_labels, labels, stats, centroids = cv2.connectedComponentsWithStats(mask_otsu_refined,
connectivity=8)

# Calculate the average size of the
# islands (excluding the background)
# and area of each label, skip label
# 0 (background):

island_sizes = stats[1:, cv2.CC_STAT_AREA]
average_island_size = np.mean(island_sizes)

# Define minimum island size as
# half the average island size:

min_island_size = average_island_size / 2

# Filter out islands smaller than the
# minimum size and keep their centroids:
valid_islands = [(size, centroid) for size, centroid in zip(island_sizes, centroids[1:]) if size >=
min_island_size]
valid_island_count = len(valid_islands)

# Calculate distances
# between neighboring centroids:

centroids_filtered = np.array([centroid for _, centroid in valid_islands])
distances = distance.pdist(centroids_filtered)     # Pairwise distances.
distance_matrix = distance.squareform(distances)  # Distance matrix.

# Measure the average and standard
# deviation of the nearest neighbor distances:

neighbor_distances = []
# Find the nearest neighbor, exclude zero distance (itself):
for i, row in enumerate(distance_matrix):
    nearest_dist = np.partition(row[row > 0], 0)[0]
    neighbor_distances.append(nearest_dist)

average_distance = np.mean(neighbor_distances)
std_dev_distance = np.std(neighbor_distances)

# Output:
print(f"Number of valid white islands (nuclei of cells) in the image: {valid_island_count}")
print(f"Average island size: {average_island_size:.2f} pixels")
print(f"Minimum island size considered (average size / 2): {min_island_size:.2f} pixels")
print(f"Average distance between neighboring nuclei: {average_distance:.2f} pixels")
print(f"Standard deviation of distances: {std_dev_distance:.2f} pixels")

plt.figure(figsize=(6, 6))
plt.imshow(mask_otsu_refined, cmap='gray')
plt.title(f'Refined Otsu Thresholding Mask - Valid Cell Count: {valid_island_count}')
plt.axis('on')
plt.show()
```

Python implementation 23: Quantitative Analysis of Nuclei Size and Distribution.

The new addition in this code is the measurement of distances between the centroids of valid nuclei (Python Implementation 6.3). After filtering, the centroids of the remaining nuclei are calculated and stored. Using these centroids, the code computes pairwise Euclidean distances between all nuclei using scipy.spatial.distance.pdist and then converts them into a distance matrix with distance.squareform. From this matrix, it calculates the nearest neighbor distance for each nucleus, excluding itself, to focus on the closest neighboring nucleus. The code then computes the average and standard deviation of these nearest neighbor distances. These distance measurements are significant because they provide insights into the spatial organization of cells within a tissue. For example, a lower average distance with a small standard deviation might indicate densely packed cells, which could be a sign of hyperplasia or neoplastic growth. Conversely, larger distances with higher variability could suggest more loosely organized or heterogeneous tissue structures, as seen in certain inflammatory conditions or benign tissues. Such spatial metrics can be used to differentiate between different tissue types or pathological states, making this approach highly valuable in veterinary and medical pathology. Therefore, the code provides a more comprehensive analysis of the tissue sample by calculating both the average size of the nuclei and the distances between them.

Output

```
Number of valid white islands (nuclei of cells) in the image: 551
Average island size: 1739.47 pixels
Minimum island size considered (average size / 2): 869.73 pixels
Average distance between neighboring nuclei: 70.50 pixels
Standard deviation of distances: 11.83 pixels
```

Next, it combines quantitative data on cell density and size with spatial distribution, giving a fuller picture of the microenvironment of the tissue. This enhanced analysis is particularly useful for identifying patterns associated with specific diseases, understanding the effects of therapeutic interventions, or conducting research studies that require detailed quantitative and spatial analysis. Note that the output of this code provides a range of informative metrics: the number of valid nuclei, the average size of these nuclei, the minimum size considered valid, and the average and standard deviation of the distances between neighboring nuclei. These results may be valuable for researchers and pathologists who need to quantify and interpret cell populations and their arrangements in tissue samples.

Morphological Parameters in Detail

The measured parameters in the output provide a detailed quantitative assessment of the nuclei within the histological image. As mentioned above, these parameters offer insights into both the size and spatial organization of the cell nuclei, which are critical for understanding tissue characteristics and potential pathological changes. This value of *number of valid white islands* (nuclei of cells), measured at a value of 551, indicates the total number of detected nuclei in the image after filtering out small, irrelevant regions. It reflects the cell density within the tissue section. A higher count may suggest hyperplasia or regions of high cellular proliferation, while a lower count might indicate atrophy, necrosis, or other conditions where cell numbers are reduced. Also, the average size of the detected nuclei is 1739.47 pixels, providing an indication of the typical size of cell nuclei within the sample. This metric is useful for identifying changes in cell size that can be indicative of specific conditions. For example, larger nuclei might suggest hypertrophy, while smaller nuclei may be associated with apoptotic or degenerative processes. Consistency in nuclear size is often a hallmark of healthy tissue, whereas significant variability could indicate dysplasia or other abnormalities. The *minimum island size considered*, evaluated at 869.73 pixels, shows the threshold used to exclude smaller objects that do not represent valid nuclei. In other words, Python Implementation 6.3 effectively filters out artifacts and noise that could skew the cell count and subsequent analyses by setting a minimum size of half the average island size. This thresholding is essential for ensuring that only biologically relevant structures are included, improving the accuracy of the analysis. Furthermore, the average distance between the centroids of neighboring nuclei is 70.50 pixels, which provides information about the spatial distribution of cells within the tissue. This metric is important for assessing tissue architecture. For instance, tightly packed nuclei with shorter average distances might indicate dense cell populations, such as in tumors or areas of inflammation. Conversely, larger distances could reflect loose or damaged tissue, as seen in certain degenerative diseases or after treatment effects. The standard deviation of 11.83 pixels for the distances between neighboring nuclei reflects the variability in cell spacing. A low standard deviation suggests that the cells are uniformly distributed, which is typically seen in organized and healthy tissues. A higher standard deviation could indicate irregular cell distribution, which may be a sign of tissue remodeling, heterogeneity, or pathological changes like neoplasia, where cells might be unevenly spaced due to abnormal growth patterns.

6.3.2 Micrometer-Scale Quantitative Analysis of Nuclei

This example takes the analysis further by converting all measurements from pixels to micrometers, providing a more biologically relevant interpretation of the histological image [91]. While it retains the same steps for image preprocessing, thresholding, and refinement as the previous codes, the significant difference lies in the use of a conversion factor to translate pixel measurements into actual physical dimensions [91]. This approach is particularly valuable in histopathology, where measurements in micrometers can offer more meaningful insights into cellular structures and their spatial relationships. The source code from Python Implementation 6.4 begins by loading the histological image and converting it to grayscale, which simplifies subsequent processing. *Otsu* thresholding is applied to separate the nuclei (white regions) from the background (black), and morphological operations are used to refine the segmentation by removing small objects and filling holes, ensuring that the detected nuclei are well defined. Next, the code then identifies distinct nuclei as "white islands" using connected component analysis and calculates the area of each island. Thus, it filters out smaller and/or irrelevant regions by determining the average size of the islands and defining a minimum size threshold at half the average, ensuring that only significant structures are considered valid nuclei.

Micrometer-Scale Quantitative Analysis of Nuclei

```
import cv2
import numpy as np
import matplotlib.pyplot as plt
from skimage.io import imread
from skimage.filters import threshold_otsu
from skimage.morphology import remove_small_objects, remove_small_holes, closing, square
from scipy.spatial import distance

image_path = r'C:\Users\Elitebook\Desktop\his_img.jpg'
image = imread(image_path)
gray_image = cv2.cvtColor(image, cv2.COLOR_RGB2GRAY)  # grayscale.

# Otsu thresholding.
otsu_thresh = threshold_otsu(gray_image)
mask_otsu = (gray_image < otsu_thresh).astype(np.uint8) * 255

# Refine the mask with
# morphological operations.
mask_otsu_refined = remove_small_objects(mask_otsu.astype(bool), min_size=64).astype(np.uint8) * 255
mask_otsu_refined = remove_small_holes(mask_otsu_refined.astype(bool),
area_threshold=64).astype(np.uint8) * 255              # Fill small holes.
mask_otsu_refined = closing(mask_otsu_refined, square(3))  # Smooth the mask.

# Count the white "islands" (nuclei of cells):
num_labels, labels, stats, centroids = cv2.connectedComponentsWithStats(mask_otsu_refined,
connectivity=8)

# Calculate the average size of
# the islands (excluding the background).
# Area of label, skip label 0 (background).
island_sizes = stats[1:, cv2.CC_STAT_AREA]
average_island_size = np.mean(island_sizes)

# Define minimum island size as
# half the average island size.
min_island_size = average_island_size / 2

# Filter out islands smaller than the
# minimum size and keep their centroids.
valid_islands = [(size, centroid) for size, centroid in zip(island_sizes, centroids[1:]) if size >=
min_island_size]
valid_island_count = len(valid_islands)

# Calculate average diameter
# in pixels for valid islands.
diameters = [2 * np.sqrt(size / np.pi) for size, _ in valid_islands]
average_diameter_pixels = np.mean(diameters)

# Calculate distances
# between neighboring centroids.
centroids_filtered = np.array([centroid for _, centroid in valid_islands])
distances = distance.pdist(centroids_filtered)    # Pairwise distances.
distance_matrix = distance.squareform(distances)  # Distance matrix.

# Measure the average and standard deviation
# of the nearest neighbor distances. Get the
# nearest neighbor, exclude zero distance (itself).
neighbor_distances = []
for i, row in enumerate(distance_matrix):
    nearest_dist = np.partition(row[row > 0], 0)[0]
    neighbor_distances.append(nearest_dist)

average_distance_pixels = np.mean(neighbor_distances)
std_dev_distance_pixels = np.std(neighbor_distances)

# Conversion factor from pixels to micrometers.
micrometers_per_pixel = 20 / 180  # ~ 0.1111 micrometers per pixel.

# Convert all measurements to micrometers.
average_island_size_um = average_island_size * (micrometers_per_pixel ** 2)
min_island_size_um = min_island_size * (micrometers_per_pixel ** 2)
average_diameter_um = average_diameter_pixels * micrometers_per_pixel
average_distance_um = average_distance_pixels * micrometers_per_pixel
std_dev_distance_um = std_dev_distance_pixels * micrometers_per_pixel

# Output results under the figure.
print(f"Number of valid white islands (nuclei of cells) in the image: {valid_island_count}")
print(f"Average island size: {average_island_size_um:.2f} micrometers^2")
print(f"Minimum island size considered (avg. size / 2): {min_island_size_um:.2f} micrometers^2")
print(f"Average island diameter: {average_diameter_um:.2f} micrometers")
print(f"Average distance between neighboring nuclei: {average_distance_um:.2f} micrometers")
print(f"Standard deviation of distances: {std_dev_distance_um:.2f} micrometers")

# Show the refined mask.
plt.figure(figsize=(8, 8))
plt.imshow(mask_otsu_refined, cmap='gray')
plt.title(f'Refined Otsu Thresholding Mask - Valid Cell Count: {valid_island_count}')
plt.axis('on')
plt.show()
```

Python implementation 24: Micrometer-Scale Quantitative Analysis of Nuclei.

What sets Python Implementation 6.4 apart from the previous versions is the introduction of calculations in micrometers. After identifying valid nuclei and computing distances between them, the code converts these measurements from pixels to micrometers using a conversion factor of approximately 0.1111 µm per pixel (derived from a scale bar of 20 µm equaling 180 pixels). This conversion is crucial because it provides a scale that relates directly to the physical world, allowing researchers and pathologists to better understand the size and spatial distribution of nuclei in a biological context. The output in micrometers gives several key parameters.

Output

```
Number of valid white islands (nuclei of cells) in the image: 551
Average island size: 21.47 micrometers^2
Minimum island size considered (average size / 2): 10.74 micrometers^2
Average island diameter: 5.82 micrometers
Average distance between neighboring nuclei: 7.83 micrometers
Standard deviation of distances: 1.31 micrometers
```

The number of valid nuclei (551 in this particular experiment) provides a measure of cell density within the tissue section, while the average island size of 21.47 μm^2 reflects the typical area of these nuclei. The minimum size considered valid is set at 10.74 μm^2, effectively filtering out noise and ensuring that the analysis focuses only on relevant cellular structures. The average island diameter of 5.82 µm provides insight into the typical size of the nuclei, which is valuable for identifying conditions such as hypertrophy or other cellular abnormalities. The spatial analysis is further enhanced by measuring the average distance between neighboring nuclei (7.83 µm) and the standard deviation of these distances (1.31 µm). The average distance indicates how closely packed the cells are within the tissue, which can be indicative of different pathological states. For instance, tightly packed cells with shorter distances between nuclei might suggest high cellularity, as seen in tumors, while more dispersed cells might indicate normal or degenerative tissue. The standard deviation provides information on the variability in spacing, with a lower value indicating uniform distribution and a higher value suggesting heterogeneity, which could be relevant in assessing tissue organization or identifying regions of necrosis, inflammation, or abnormal growth. Note that since the same input image was used, Python Implementation 6.4 produces the same figure as in the previous example (Fig. 6.2).

6.4 Pixel-to-Micrometer Conversion in Histological Image Analysis

In histological image analysis, the accurate interpretation of tissue architecture and cellular structures depends not only on the quality of the images but also on the understanding of scales [135]. Scales, such as bar scales and magnification levels, provide crucial context that translates digital measurements from pixel-based data to meaningful biological dimensions [91]. These conversions are essential for making quantitative assessments that are consistent, reproducible, and comparable across different studies and clinical settings [135]. The ***bar scale*** is a fundamental reference tool found on most histological images. It indicates the real-world distance that corresponds to a certain number of pixels on the image. For instance, a bar scale might show that a segment of 180 pixels represents 20 μm in actual size. The bar scale allows for the conversion of pixel measurements obtained from digital image analysis into micrometers or other real-world units, thus providing a clear visual reference. This is essential for any form of quantitative analysis, such as measuring cell sizes, distances between cells, or the area of tissue regions. Without the bar scale, any pixel-based measurement would lack a meaningful context, rendering it useless for biological interpretation [135]. Magnification is another critical aspect that directly influences the interpretation of histological images. The magnification level of the microscope determines how much a particular area of a tissue sample is enlarged in the digital image. For example, a 40× magnification means the image is 40 times larger than the actual size of the tissue being observed. Understanding magnification is vital when assessing cellular features, as it affects the scale and resolution of the captured image. High magnification provides detailed views of small structures, such as cell nuclei, while lower magnification offers a broader overview of the tissue architecture. Knowing the magnification used for an image allows researchers to accurately correlate the observed structures to their actual sizes, enabling more precise analysis. The conversion of pixels to micrometers is a process that ties together the concepts of the bar scale and magnification to provide a quantifiable measure of size and distance in histological images. This conversion is essential for translating the pixel-based measurements generated by digital image analysis tools into real-world units that are biologically meaningful. For example, when a bar scale shows that 180 pixels equal 20 μm, the conversion factor can be derived as approximately 0.1111 μm per pixel. This conversion factor is then used to transform all pixel-based measurements, such as cell size, diameter, or distance between nuclei, into micrometers. The conversion of these measurements to micrometers is particularly important for comparing findings across different studies, where magnification levels and image resolutions may vary. It allows for a standardized interpretation of results, making it possible to draw meaningful biological conclusions. Accurate scaling and conversion are vital in histopathology for several reasons. They allow pathologists and researchers to quantify cell sizes, densities, and spatial distributions, which are important indicators of tissue health and disease states. For instance, variations in nuclear size and shape can be indicative of malignancy or other pathological conditions. Similarly, the distance between neighboring nuclei can provide insights into tissue organization, cell proliferation rates, or areas of necrosis. Without accurate scale references, such measurements would lack context, leading to incorrect or inconsistent conclusions. Moreover, the use of scales and conversions extends beyond the basic measurements. It is crucial for advanced quantitative image analysis methods that involve machine learning and artificial intelligence. For these models to be effective and generalizable, the input data must be accurately scaled to reflect real-world dimensions. Properly scaled data ensures that the outputs of the model are biologically relevant and can be used to make reliable clinical decisions.

6.5 Quantifying Staining Intensity and Distribution

Quantifying staining intensity and distribution is a critical aspect of histological analysis in veterinary pathology, providing essential information about the presence, abundance, and localization of specific cellular or extracellular components. Staining techniques, such as hematoxylin and eosin (H&E) or immunohistochemistry (IHC), are widely used to highlight various tissue elements, making them visible under a microscope. However, traditional visual assessments of these stained sections can be subjective and prone to variability between observers. Quantitative image analysis addresses these limitations by providing objective and reproducible measurements of staining intensity and distribution. Advanced image analysis software can quantify staining intensity by measuring the pixel values within regions of interest, reflecting the concentration of the stain. For example, darker or more intense staining might indicate a higher concentration of a particular protein or cellular component, which could be relevant for diagnosing conditions like inflammation, neoplasia, or metabolic diseases. The software implementations can accurately measure intensity levels by simply converting color images into grayscale or other color spaces that separate hue, saturation, and brightness, even in complex tissues where multiple stains are used simultaneously. Quantification of the distribution of staining is equally important in veterinary pathology, as it reveals patterns that can be diagnostic or prognostic. Analysis tools can map the spatial distribution of staining across an entire tissue section, providing detailed information on how a particular marker is expressed in different areas. This can be especially useful for identifying regions of pathological significance, such as tumor margins, areas of necrosis, or sites of immune cell infiltration. Also, pathologists can better understand disease progression and response to therapy by using different algorithms to segment and analyze these regions. One of the most powerful aspects of quantifying staining intensity and distribution is the ability to correlate these measurements with clinical outcomes. In research settings, for instance, quantitative analysis of IHC staining can help establish relationships between marker expression levels and disease stages or treatment responses. This approach allows for the development of more accurate prognostic models and personalized treatment plans, ultimately improving patient outcomes. In a veterinary context, such quantification is invaluable for understanding how diseases manifest differently across species and developing species-specific diagnostic and therapeutic strategies. Automated image analysis also enables high-throughput screening of histological samples, which is particularly advantageous in large-scale studies or clinical trials. The ability to process and analyze vast numbers of slides quickly ensures that results are consistent and not influenced by individual bias. Moreover, these techniques can be integrated with machine learning algorithms to refine and improve the accuracy of staining quantification over time, making them an indispensable part of modern veterinary diagnostics and research.

6.5.1 Staining Intensity and Distribution

In synchronicity with the above, the next code helps in the field by providing a way to visualize and quantify staining intensity distributions in histological images (Python Implementation 6.5). The histogram visualization guides users in selecting appropriate thresholds for segmenting different intensity regions, allowing for better differentiation of tissue types or identifying areas of interest. This analysis can be useful for quality control in staining procedures, detection of uneven staining or artifacts, and assertion of the presence and distribution of specific stains or biomarkers that may be relevant in research or diagnostic contexts. Thus, Python Implementation 6.6 builds upon the previous examples of quantifying staining intensity in histological images by adding a step to visualize the intensity distribution of the pixels over the image. It uses the value (V) channel from the HSV color space to analyze the intensity of the image, making it useful for understanding the distribution of staining intensities across a tissue sample. The primary purpose of this code is to provide a way to visually assess the overall staining pattern and to segment regions of interest based on their intensity values. In the context of the previous codes, which focused on cell counting, size measurements, and spatial distributions, this code shifts the focus toward understanding the intensity profile of the stains. This is particularly valuable for pathologists and researchers who need to assess the quality and uniformity of staining or to identify areas of different staining intensity, which could correlate with distinct tissue types, levels of cellular activity, or presence of specific biomarkers. The implementation starts by loading a histological image and converting it to the hue, saturation, value (HSV) color space. The HSV model is preferred over RGB for intensity analysis because it separates the intensity information (value) from color information (hue and saturation). This separation allows for a more focused analysis of staining intensity without interference from color variations. The value (V) channel represents the brightness or intensity of each pixel, making it the ideal choice for quantifying staining intensity. Next, the code generates a histogram of the intensity values in the value channel. This histogram is visualized in a figure to show the distribution of pixel intensities across the entire image. The histogram displays the frequency of intensity values ranging from 0 to 1 (normalized intensity scale). The x-axis represents the intensity values, while the y-axis shows the frequency (number of pixels) corresponding to each intensity level. The histogram is essential for understanding the overall contrast and brightness of the image. In a typical histological image, the histogram might show peaks corresponding to regions with high staining (dark areas) and lower staining (lighter areas). If the staining is highly uniform or if the image has a narrow range of intensities, the histogram would show a concentrated peak within a specific range.

Staining intensity and distribution

```
import cv2
import numpy as np
import matplotlib.pyplot as plt
from skimage.io import imread
from skimage.color import rgb2hsv

image_path = r'C:\Users\Elitebook\Desktop\his_img.jpg'
image = imread(image_path)

# Convert the image to HSV color space.
hsv_image = rgb2hsv(image)

# Extract the Value (V) channel,
# which represents intensity.
intensity_channel = hsv_image[:, :, 2]

# Visualize the intensity
# distribution using a histogram.
plt.figure(figsize=(8, 6))
plt.hist(intensity_channel.ravel(), bins=256, range=(0, 1), color='black', alpha=0.7)
plt.title('Intensity Distribution (Value Channel)')
plt.xlabel('Intensity')
plt.ylabel('Frequency')
plt.show()

# Thresholds:
low_threshold = 0.8
high_threshold = 0.8

# Thresholding to segment regions with
# different staining intensity levels.
low_intensity_mask = intensity_channel < low_threshold
high_intensity_mask = intensity_channel > high_threshold

# Visualize the original image with
# segmented low and high-intensity regions.
plt.figure(figsize=(15, 5))

plt.subplot(1, 3, 1)
plt.imshow(image)
plt.title('Original Image')
plt.axis('on')

plt.subplot(1, 3, 2)
plt.imshow(low_intensity_mask, cmap='gray')
plt.title(f'Low Intensity Regions (< {low_threshold})')
plt.axis('on')

plt.subplot(1, 3, 3)
plt.imshow(high_intensity_mask, cmap='gray')
plt.title(f'High Intensity Regions (> {high_threshold})')
plt.axis('on')

plt.show()
```

Python implementation 25: Staining intensity and distribution.

The intensity histogram can look different depending on the staining technique used and the tissue type. For example, an H&E-stained histological section might show two distinct peaks: one for the lighter background and cytoplasm (lower-intensity values) and another for the darker nuclei (higher-intensity values). If the histogram appears to be skewed heavily toward one side, it suggests that the image is dominated by either very light or very dark regions, which could indicate over-staining, under-staining, or uneven staining (Fig. 6.3). Based on the analysis of the intensity histogram, the code sets thresholds for segmenting regions of different staining intensities. The *low_threshold* and *high_threshold* parameters are both set to 0.8, indicating a specific intensity level chosen for

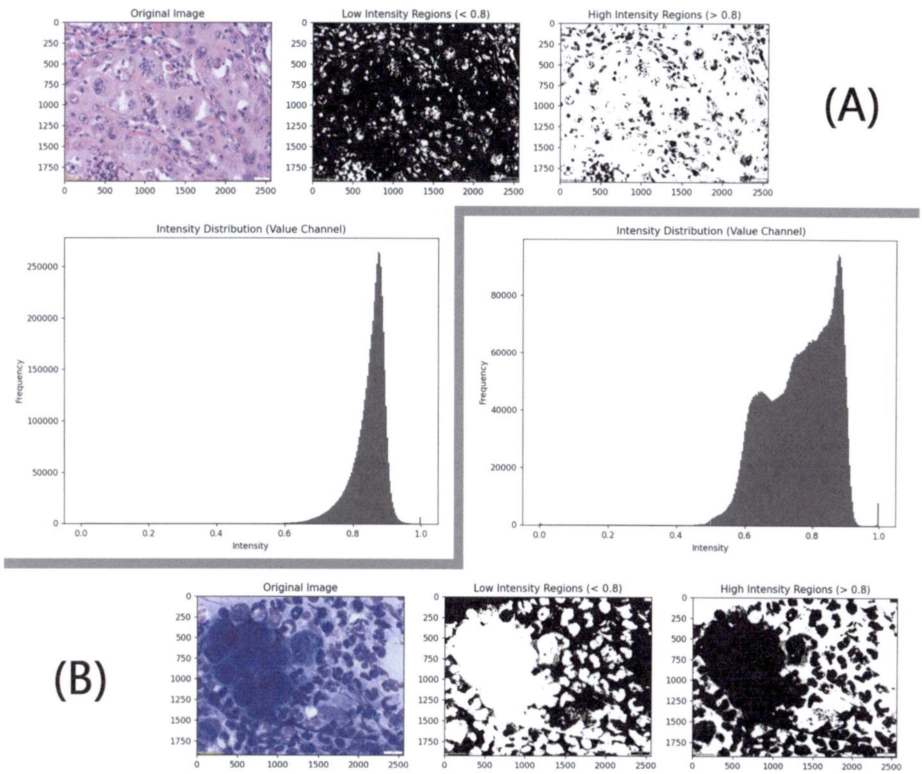

Fig. 6.3 Staining intensity and distribution analysis on two separate histological images. Panel (**a**) shows a histological section of a malignant feline skin tumor diagnosed as squamous cell carcinoma, captured at 40× magnification with a scale bar representing 20 μm. On the other hand, panel (**b**) depicts cells from a feline lower lip lesion, captured at 100× magnification with a scale bar representing 10 μm. The intensity histograms of these images illustrate the distribution of staining intensity across the tissue sections. Variations in histogram profiles reflect differences in staining techniques and tissue types. For instance, in hematoxylin and eosin (H&E) staining, two distinct peaks might be seen: one representing lighter cytoplasmic regions and another for darker nuclei. Skewness in the histogram could indicate issues like over-staining, under-staining, or uneven staining, as observed in these panels. Based on the histogram analysis, thresholds are set to segment regions of different staining intensities. Both low and high thresholds are set to 0.8, capturing only the most intense regions of the tissue. In the low-intensity mask, darker areas would be highlighted, but due to the high threshold, the mask may appear mostly black, indicating minimal regions below the threshold. Conversely, the high-intensity mask isolates the brightest regions exceeding the threshold, likely resulting in sparse segmentation. The choice of threshold impacts the visibility of tissue components, and in these cases, the high threshold may only capture the most distinct regions

segmentation. These thresholds are used to create binary masks that highlight regions with intensities below or above the chosen value. The *low_intensity_mask* captures all regions where the intensity is below 0.8, while the *high_intensity_mask* captures regions where the intensity is above 0.8.

The segmented regions are then visualized in a series of subplots: the original image, the low-intensity mask, and the high-intensity mask (Fig. 6.3a, b). In the low-intensity mask, areas that are darker (less stained or have low intensity) would be highlighted if the threshold is appropriately set. Since the threshold in this example is set to 0.8, which is quite high, the low-intensity mask will likely appear black for some images, indicating that most areas do not fall below this high threshold. Conversely, the high-intensity mask will show the brightest areas of the image that exceed the threshold of 0.8. If the image has very few areas of such high intensity, only a few scattered spots will be visible in the high-intensity mask. The choice of a high threshold value (0.8) might not be optimal for most histological images unless the goal is to specifically isolate only the very brightest regions. Typically, more balanced thresholds (such as 0.3 for low intensity and 0.7 for high intensity) may be used to capture a broader range of staining intensity and provide a more meaningful segmentation of tissue components.

6.5.2 Description of Results

The results presented above show a comparison of staining intensity and distribution in two different histological images using thresholds to segment low- and high-intensity regions (Fig. 6.3a, b). The analysis relies on the *value* (V) channel from the HSV color space to identify regions with different staining intensities, which are relevant for assessing tissue architecture, cellular composition, and quality of staining in histopathological studies. The first panels at the top of the figure show the analysis for one histological image (Fig. 6.3a). The original image shows densely stained tissue with darkly stained regions. The *lowintensity regions* (<0.8) mask highlights the areas where the intensity values are below 0.8. In this image, most regions have relatively high intensity, indicating dense staining with a significant amount of dark regions. The *high-intensity regions* (>0.8) mask shows areas where the intensity is above 0.8, revealing regions with very light staining. Due to the high threshold value of 0.8, most areas in this mask are white, showing only a few scattered spots of high intensity (Fig. 6.3a). The segmentation here is based on separating regions with light and dark staining, allowing for an assessment of areas where staining is either too intense or lacking. The panel below the original image represents the histogram for the intensity distribution of the *value* (V) channel for this first image (Fig. 6.3a). The histogram has a sharp peak around 0.8–0.9, suggesting that most pixel values are concentrated in the high-intensity range. This indicates that the image contains mostly bright regions with some very dark areas, likely due to dense staining or heavy dye application. This is typical of highly stained histological samples, where certain regions absorb more dye than others. The next set of panels shows the analysis for a different histological image (Fig. 6.3b). The original image appears less densely stained than the previous one, with more balanced staining across the tissue. The *low-intensity regions* (<0.8) mask for this image reveals a different pattern compared to the first image, where more regions fall below the intensity threshold of 0.8, indicating that there are more low-intensity areas (less densely stained regions). The *high-intensity regions* (>0.8) mask displays more areas than the first image, suggesting that more regions in this sample have high intensity, potentially highlighting areas of lighter staining or regions with more transparent tissues (Fig. 6.3b). The second histogram (on the right) shows the intensity distribution for this second image (Fig. 6.3b). It is more spread out with a broader peak ranging from around 0.6–1.0. This suggests a more varied range of staining intensities, with more regions of both low and high intensities. Unlike the first image, where the intensity distribution is heavily skewed toward high-intensity values, this image shows a more balanced distribution, which may indicate better staining uniformity or different tissue properties.

6.5.3 Interpretation and Importance in Comparative Analysis

A comparison of these two images provides valuable insights into the differences in staining intensity and distribution between them (Fig. 6.3a, b). The first image, with its skewed histogram and concentrated high-intensity regions, may suggest either over-staining or a tissue type that tends to absorb more stain (Fig. 6.3a). The presence of fewer low-intensity regions implies that the staining process might have heavily targeted specific tissue components, possibly highlighting nuclear regions or other dense structures. The second image shows a more balanced intensity distribution, with both low- and high-intensity regions more evenly represented (Fig. 6.3b). This can indicate a more evenly stained sample, where various tissue components are more uniformly highlighted. This difference could be due to different staining techniques, varying dye concentrations, or differences in the tissue types themselves. Please notice that these analyses can be useful in the field of pathology and veterinary medicine as they help in assessing the quality and effectiveness of staining protocols or comparing tissue characteristics.

6.6 Statistical Analysis of Quantitative Data

Statistical analysis of quantitative data is a fundamental component in veterinary pathology, providing the tools needed to make sense of complex datasets derived from quantitative image analysis [144]. In the context of veterinary medicine, histological images, cell counts, morphometric measurements, and other quantitative parameters are often collected in large volumes [145]. In order to extract meaningful insights from this data, robust statistical methods must be employed to evaluate patterns, relationships, and differences among various samples, conditions, or treatment groups [146]. Descriptive statistics play a crucial role in summarizing the data by providing measures such as means, medians, standard deviations, and ranges. These statistics give a snapshot of the central tendencies and variability within the dataset, which is essential for understanding the general characteristics of the biological samples under study [147]. For instance, when analyzing cell sizes or densities in different tissue types or pathological conditions, descriptive statistics can highlight differences that might suggest disease presence or progression [134, 148]. Beyond basic summaries, inferential statistics allow veterinarians and researchers to draw conclusions about larger populations based on sample data [145]. Techniques such as hypothesis testing, analysis of variance (ANOVA), and regression analysis are commonly used to determine whether observed differences between groups are statistically significant or likely to have occurred by chance. In veterinary research, this could mean determining whether a particular treatment significantly reduces tumor size or alters cellular morphology compared to a control group. These methods help to establish causal relationships and validate findings, thereby providing stronger evidence for clinical or therapeutic decisions. *Multivariate statistical methods* are particularly useful when dealing with complex datasets, where multiple variables may interact or influence each other [149]. Techniques like principal component analysis (PCA) and cluster analysis can be employed to reduce dimensionality, identify patterns, or group similar samples [149, 150]. This is especially valuable in veterinary pathology when dealing with high-dimensional data such as those generated from high-throughput imaging or molecular studies. Because the complexity of the data is much reduced, multivariate methods can help identify underlying trends or classifications that are not immediately apparent from univariate analysis [151]. Correlation and regression analyses are also pivotal in exploring relationships between variables. For example, regression models can predict outcomes such as disease progression or response to treatment based on quantitative image features like cell density, size, or spatial arrangement. These models provide a framework for understanding how different histological features relate to clinical outcomes, enabling more targeted and effective therapeutic strategies. Advanced statistical methods, including machine learning–based approaches, are increasingly being integrated into quantitative image analysis [10]. These methods, such as random forests, support vector machines, and neural networks, can handle large, complex datasets and are particularly adept at identifying nonlinear relationships and interactions between variables [149]. Machine learning models can be trained to recognize specific pathological features or predict disease outcomes based on quantitative data, further enhancing the diagnostic power of quantitative image analysis in veterinary pathology. The choice of statistical methods depends on the nature of the data, the specific research questions, and the desired outcomes [84, 148]. Proper statistical analysis ensures that the conclusions drawn from quantitative data are valid, reliable, and meaningful. It requires careful planning, including considerations for sample size, data distribution, and potential confounding factors, to avoid the pitfalls such as overfitting or false positives.

6.7 Identification of Statistical Parameters

Python Implementation 6.6 builds on the concepts introduced in the earlier examples by incorporating statistical analysis and visualization of quantitative data obtained from histological image analysis. While the previous codes focused on identifying and measuring cell nuclei (nuclei detection, cell counting, size calculation, and distance measurement), this example takes it a step further by statistically analyzing the data, which provides a deeper understanding of the tissue characteristics and patterns. The implementation generates descriptive statistics, performs hypothesis testing, calculates correlations, and visualizes the data distributions through histograms and scatter plots. It uses libraries such as *numpy*, *pandas*, *scipy*, and *matplotlib* to perform these tasks, which are vital for data interpretation in a scientific context. Note that to prevent any issue regarding the execution of the code, one is advised to install *pandas* and any other libraries that might not be installed, such as *numpy*, *matplotlib*, and *scipy*. Thus, in the command prompt or terminal, run the following commands:

Console

```
pip install pandas
pip install numpy
pip install matplotlib
pip install scipy
```

Once these libraries are installed, the code should run without problems (Python Implementation 6.6). To continue, the use of descriptive statistics provides summary metrics like mean, standard deviation, and range, giving an overview of the variability and the central tendency of the data. Hypothesis testing checks whether the mean cell sizes and distances are significantly different from a given value, allowing for comparisons against known or expected benchmarks. The correlation analysis between cell size and neighbor distance provides insights into the potential relationships between these two variables, which could indicate specific tissue or pathological conditions. But what do the charts produced by this code mean?

Statistical analysis of quantitative data

```
import numpy as np
import pandas as pd
import matplotlib.pyplot as plt
from scipy import stats

# Sample data representing measurements from histological
# analysis. Cell size (in micrometers^2) and distances
# between neighboring nuclei (in micrometers).
cell_sizes = np.random.normal(loc=20, scale=5, size=100)          # Simulated cell sizes.
neighbor_distances = np.random.normal(loc=7, scale=1.5, size=100)  # Distances between nuclei.

# Convert data to a DataFrame for better handling:
data = pd.DataFrame({
    'Cell Size (micrometers^2)': cell_sizes,
    'Neighbor Distance (micrometers)': neighbor_distances})

# Descriptive Statistics.
print("Descriptive Statistics for Cell Sizes:")
print(data['Cell Size (micrometers^2)'].describe())
print("\nDescriptive Statistics for Neighbor Distances:")
print(data['Neighbor Distance (micrometers)'].describe())

plt.figure(figsize=(12, 5)) # Visualize data distributions.
plt.subplot(1, 2, 1)
plt.hist(cell_sizes, bins=15, color='skyblue', edgecolor='black')
plt.title('Histogram of Cell Sizes')
plt.xlabel('Cell Size (micrometers^2)')
plt.ylabel('Frequency')

plt.subplot(1, 2, 2)
plt.hist(neighbor_distances, bins=15, color='salmon', edgecolor='black')
plt.title('Histogram of Neighbor Distances')
plt.xlabel('Distance Between Nuclei (micrometers)')
plt.ylabel('Frequency')
plt.tight_layout()
plt.show()

# Hypothesis Testing: Are the mean cell sizes and
# neighbor distances significantly different from
# a given value? Test if mean cell size is 20
# micrometers^2. Also, test if mean neighbor distance is 7 micrometers:
t_stat_size, p_val_size = stats.ttest_1samp(cell_sizes, 20)
t_stat_dist, p_val_dist = stats.ttest_1samp(neighbor_distances, 7)

print(f"\nT-test - mean cell size: t-stat. = {t_stat_size:.2f}, p-value = {p_val_size:.4f}")
print(f"T-test - mean neighbor distance: t-stat. = {t_stat_dist:.2f}, p-value = {p_val_dist:.4f}")

# Correlation Analysis: Is there a correlation
# between cell size and neighbor distance?
correlation, p_value = stats.pearsonr(data['Cell Size (micrometers^2)'], data['Neighbor Distance
(micrometers)'])
print(f"\nCorrelation between Cell Size and Neighbor Distance: Pearson correlation =
{correlation:.2f}, p-value = {p_value:.4f}")

# Visualize the correlation with a scatter plot:
plt.figure(figsize=(6, 6))
plt.scatter(data['Cell Size'], data['Neighbor Distance'], alpha=0.7, color='purple')
plt.title('Scatter Plot of Cell Size vs. Neighbor Distance')
plt.xlabel('Cell Size (micrometers^2)')
plt.ylabel('Neighbor Distance (micrometers)')
plt.grid(True)
plt.show()
```

Python implementation 26: Statistical analysis of quantitative data.

The left-bottom panel from the main figure is a scatter plot that shows the relationship between cell size (in μm^2) and neighbor distance (in μm). The scatter plot indicates that there is no strong correlation between the two variables, as the points are widely spread without forming a clear linear pattern (Fig. 6.4). This suggests that the size of the cells (nuclei) and the distance between them are relatively independent of each other in this

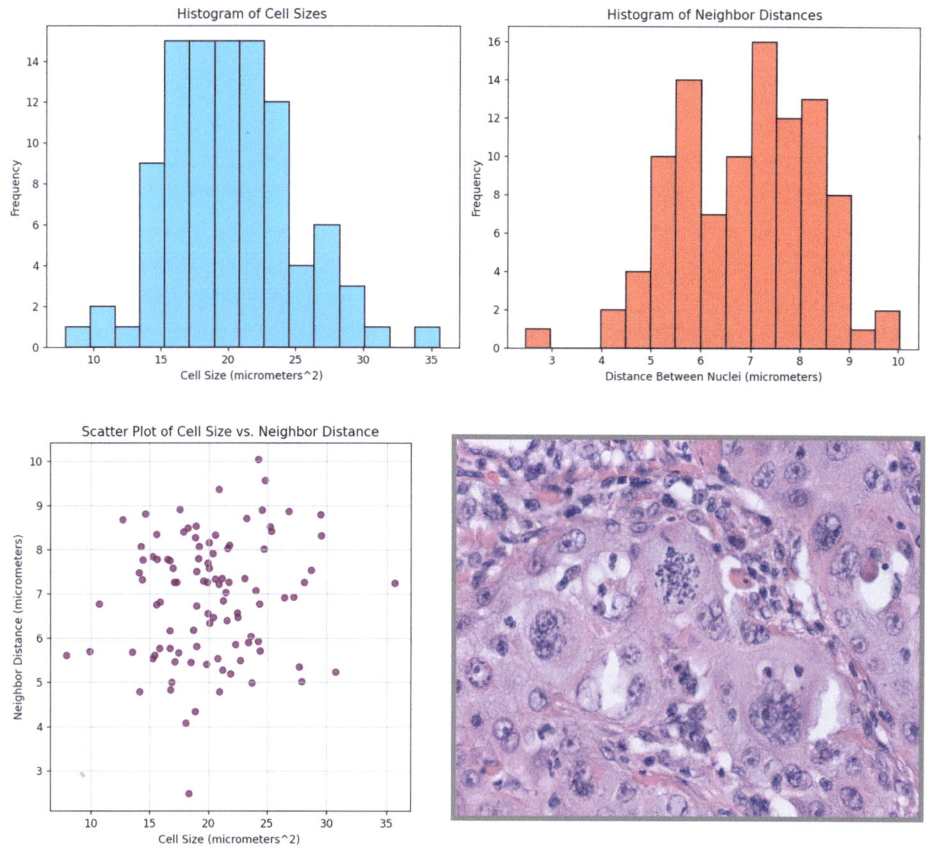

Fig. 6.4 This figure presents a quantitative analysis of an image representing squamous cell carcinoma (SCC) in the skin of a cat. The panels display both statistical and visual data to explore the cellular morphology and spatial distribution of nuclei within the tissue. The top-left panel shows a histogram of cell sizes (in μm^2), which provides a frequency distribution of measured cell sizes within the carcinoma tissue. The data reveal a relatively normal distribution, with most cells falling between 15 and 25 μm^2, indicating a range of cell sizes typical for this type of tissue. The top-right panel illustrates a histogram of neighbor distances between nuclei, measured in micrometers. This histogram indicates the frequency of distances between adjacent nuclei, with a distribution peak between 6 and 8 μm. This information is critical for understanding the spatial arrangement of cells and detecting abnormalities in nuclear distribution, which can be indicative of malignancy. The bottom-left panel presents a scatter plot of cell size versus neighbor distance, where each point represents an individual cell. The x-axis indicates the cell size in μm^2, while the y-axis represents the distance between the nucleus of that cell and its nearest neighbor. The scatter plot helps in identifying any potential correlation between cell size and spacing, which could suggest cellular behaviors or abnormalities in the carcinoma context. The bottom-right panel shows the original histological image of the squamous cell carcinoma tissue from the cat (imaged at 40× magnification with a scale bar of 20 μm). The image reveals the cellular architecture and morphological details, including densely packed clusters of cells with irregular nuclei—a characteristic feature of squamous cell carcinoma. This visual data is crucial for understanding the morphological context that supports the quantitative analysis displayed in the other panels. Together, these panels provide a comprehensive overview of both the quantitative and qualitative characteristics of squamous cell carcinoma in the feline skin, facilitating a deeper understanding of tumor morphology and potential diagnostic markers

sample. The scatter plot is useful in exploring whether larger cells are more spaced out or closely packed together. In this case, there is no evident trend, implying that cell proliferation and spacing do not necessarily correlate in the observed tissue section. The top-left and top-right panels of the figure show histograms of the cell sizes and neighbor distances, providing a visual representation of their distributions (Fig. 6.4). The histogram for cell sizes appears roughly normally distributed, centered around a mean of approximately 20 μm^2 (Fig. 6.4). This suggests that most cells in this tissue sample have a similar size, with only a few being significantly smaller or larger. Such a distribution could indicate a relatively homogeneous cell population, typical of healthy tissue or a specific tissue type. The histogram for neighbor distances shows a slightly skewed distribution, peaking around 7 μm. This indicates that most cells have a consistent spacing relative to their neighbors, which could reflect an organized tissue structure, such as a healthy epithelial layer. However, some variability is present, as indicated by the spread of the distribution, suggesting that not all cells are uniformly spaced. The right-bottom image is a histological slide stained with hematoxylin and eosin (H&E) at 40× magnification, with a scale bar indicating 20 μm (Fig. 6.4). This image serves as the raw data source for quantitative analysis, where the nuclei are prominently stained dark purple, making them easily distinguishable from the surrounding cytoplasm and extracellular matrix. The distinct staining of nuclei is due to the presence of heterochromatin, which binds the hematoxylin stain more intensely, creating a contrast that is ideal for segmentation and analysis. The varying densities and sizes of these nuclei in the image provide valuable information about the cellular architecture of the tissue. In combination with the previous figures, it becomes clear how quantitative analysis of features such as cell size and spacing can provide insights into the health and organization of the tissue, potentially aiding in the diagnosis of pathological conditions.

6.8 Description of Results

The above implementation and the attached visual results illustrate the importance of combining image processing, statistical analysis, and data visualization in histological image analysis. The ability to convert pixel measurements into biologically relevant units, such as micrometers, allows for more accurate comparisons across different samples and magnifications, ensuring the reliability of the findings. As such, researchers and pathologists can gain deeper insights into tissue organization, cellular morphology, and potential disease states by leveraging these quantitative and statistical tools. Please notice that the console output provides a comprehensive summary of the quantitative analysis of cell sizes and distances between neighboring nuclei in the histological image, along with the results of statistical tests that offer further insights into the characteristics of the tissue.

Output

```
Descriptive Statistics for Cell Sizes:
count 100.000000
mean 20.179091
std 4.688184
min 7.909224
25% 16.935933
50% 19.959652
75% 23.111630
max 35.623742
Name: Cell Size (micrometers^2)

Descriptive Statistics for Neighbor Distances:
count 100.000000
mean 6.914051
std 1.377304
min 2.477154
25% 5.747661
50% 7.152838
75% 7.936222
max 10.043523
Name: Neighbor Distance (micrometers)

T-test for mean cell size = 20 micrometers^2: t-statistic = 0.38, p-value = 0.7033
T-test for mean neighbor distance = 7 micrometers: t-statistic = -0.62, p-value = 0.5340
Correlation between Cell Size and Neighbor Distance: Pearson correlation = 0.12, p-value = 0.2232
```

The descriptive statistics for cell sizes show that the mean cell size is approximately 20.18 μm^2 (for this particular histological image), with a standard deviation of 4.69 μm^2. This indicates a moderate level of variability in cell sizes around the mean. The minimum and maximum values of cell size are 7.91 and 35.62 μm^2, respectively, suggesting a range of cell sizes from very small to relatively large within the tissue. The 25th, 50th (median), and 75th percentiles are 16.94, 19.96, and 23.11 μm^2, respectively, indicating that most cell sizes are concentrated around the mean, with a slight skew toward larger sizes. This distribution aligns with the histogram provided earlier, where the cell sizes show a roughly normal distribution with a peak near 20 μm^2. The descriptive statistics for neighbor distances reveal that the average distance between neighboring nuclei is 6.91 μm, with a standard deviation of 1.38 μm. This reflects some variation in how closely packed the nuclei are, with most cells having neighboring nuclei at distances ranging from about 5.75 μm (25th percentile) to 7.94 μm (75th percentile). The minimum and maximum neighbor distances are 2.48 and 10.04 μm, respectively, indicating that while most cells are relatively closely spaced, there are a few instances where cells are either very close together or more dispersed. The histogram of neighbor distances shows a slight skewness, supporting this observation.

The *t-tests* for the mean values of cell size and neighbor distance provide additional insights. For cell size, the t-statistic is 0.38 with a *p*-value of 0.7033, indicating no significant difference from the hypothesized mean of 20 μm^2. This result suggests that the observed mean cell size is consistent with what might be expected based on prior knowledge or a control sample. Similarly, the *t-test* for the mean neighbor distance yields a t-statistic of −0.62 and a *p*-value of 0.5340, again indicating no significant difference from the hypothesized mean of 7 μm. Both *p*-values are well above the common significance level of 0.05, suggesting that the observed sample means do not differ significantly from the hypothesized population means. Next, the correlation analysis between cell size and neighbor distance shows a *Pearson* correlation coefficient of 0.12 with a p-value of 0.2232. A correlation coefficient of 0.12 suggests a very weak positive relationship between cell size and neighbor distance, meaning that there is almost no linear association between these two variables in this dataset. The p-value of 0.2232 further indicates that this weak correlation is not statistically significant, reinforcing the visual observation from the scatter plot that the two variables are largely independent of each other. Furthermore, the results suggest a relatively homogeneous distribution of cell sizes with moderate variability in the distances between neighboring nuclei. The lack of significant correlation between cell size and neighbor distance implies that in this particular tissue sample, larger cells do not necessarily have more space around them, nor are smaller cells more tightly packed. This information can be crucial for understanding the tissue organizational structure and for identifying any deviations that may indicate pathological changes. The combination of descriptive statistics, hypothesis testing, and correlation analysis provides a robust framework for quantitatively assessing histological images, making it valuable for research, diagnostics, and treatment planning in veterinary pathology.

6.9 Applications in Disease Diagnosis and Prognosis

Quantitative image analysis has become an essential tool in the diagnosis and prognosis of diseases in veterinary medicine, providing a level of precision and consistency that surpasses traditional diagnostic methods. In the diagnosis of neoplastic diseases, such as various forms of cancer, quantitative image analysis enables the precise identification and quantification of abnormal cell growth, nuclear atypia, and mitotic activity. Through automated detection and classification algorithms, pathologists can objectively evaluate tumor grade, assess margins, and identify metastatic potential, leading to more accurate diagnoses and treatment planning. The ability to analyze multiple morphological features simultaneously—such as size, shape, and texture of cells—enhances the detection of subtle changes that may be indicative of early malignancy or progression, offering a significant advantage over conventional methods that rely heavily on subjective visual interpretation. Nevertheless, beyond oncology, quantitative image analysis is highly valuable in diagnosing infectious and inflammatory diseases. The technology can be used to quantify inflammatory cell infiltration, identify specific cell types, and evaluate tissue architecture alterations. For instance, in cases of chronic inflammatory conditions or infectious diseases caused by bacteria, fungi, or parasites, the software can measure and track changes in cell density, distribution, and interaction over time, providing critical data for understanding disease dynamics and progression. This capability not only aids in diagnosis but also allows veterinarians to monitor the effectiveness of therapeutic interventions and adjust treatments accordingly. Prognostic assessments are also significantly enhanced through quantitative image analysis. The extraction of quantitative features and the histological patterns found may help predict disease outcomes and patient survival. For example, in cases of canine lymphoma or feline mammary carcinoma, the presence of specific biomarkers and the architectural patterns of tumor cells can be quantified to provide prognostic insights. Automated image analysis tools can process large datasets to identify prognostic markers that are not readily apparent through manual examination, such as microvascular density or stromal composition, which have been linked to tumor aggressiveness and response to therapy. Moreover, quantitative image analysis can be integrated with other diagnostic modalities, such as immunohistochemistry and molecular pathology, to provide a more comprehensive view of disease pathology.

6.10 Detection of Abnormally Sized Nuclei for Malignant Tumor Identification

Computational identification of malignant tumors is the only *Holy Grail* of veterinary medicine. Therefore, a few examples are in order. The code shown in Python Implementation 6.7 is designed to detect abnormally sized nuclei within a histological image, which can be an indicator of malignancy in tissue samples. In other words, the code aims to facilitate the detection of potential malignant tumors, where cells often exhibit larger and irregularly shaped nuclei. First, as many times before, the code begins by loading a histological image and converting it to grayscale, which simplifies the segmentation process. *Otsu* thresholding is applied to separate the nuclei (foreground) from the background. This method automatically determines an optimal threshold value based on the grayscale histogram, making it suitable for images with varying staining intensities. Following thresholding, morphological operations, including the removal of small objects and the filling of small holes, refine the binary mask. This step ensures that only well-defined, significant nuclei are retained, which is crucial for accurate detection and analysis.

Detection of abnormally sized nuclei for malignant tumor identification

```
import cv2
import numpy as np
import matplotlib.pyplot as plt
from skimage.io import imread
from skimage.filters import threshold_otsu
from skimage.morphology import remove_small_objects, remove_small_holes, closing, square
from scipy.spatial import distance

image_path = r'C:\Users\Elitebook\Desktop\his_img.jpg'
image = imread(image_path)
gray_image = cv2.cvtColor(image, cv2.COLOR_RGB2GRAY)

otsu_thresh = threshold_otsu(gray_image)
mask_otsu = (gray_image < otsu_thresh).astype(np.uint8) * 255

# Refine the mask with morphological
# operations and remove small objects:

mask_otsu_refined = remove_small_objects(mask_otsu.astype(bool), min_size=64).astype(np.uint8) * 255
mask_otsu_refined = remove_small_holes(mask_otsu_refined.astype(bool),
area_threshold=64).astype(np.uint8) * 255                # Fill small holes.
mask_otsu_refined = closing(mask_otsu_refined, square(3))  # Smooth the mask.

# Count the white "islands"
# (nuclei of cells):

num_labels, labels, stats, centroids = cv2.connectedComponentsWithStats(mask_otsu_refined,
connectivity=8)

# Calculate the average size of the islands
# (excluding the background) and calc. the
# area of each label, skip label 0 (background):

island_sizes = stats[1:, cv2.CC_STAT_AREA]
average_island_size = np.mean(island_sizes)

# Define minimum island size
# as half the average island size:

min_island_size = average_island_size / 2

# Filter out islands smaller than the
# minimum size and keep their centroids:

valid_islands = [(size, centroid, i + 1) for i, (size, centroid) in enumerate(zip(island_sizes,
centroids[1:])) if size >= min_island_size]
valid_island_count = len(valid_islands)

# Calculate average diameter
# in pixels for valid islands:

diameters = [2 * np.sqrt(size / np.pi) for size, _, _ in valid_islands]
average_diameter_pixels = np.mean(diameters)

# Calculate distances between
# neighboring centroids:

centroids_filtered = np.array([centroid for _, centroid, _ in valid_islands])
distances = distance.pdist(centroids_filtered)    # Pairwise distances.
distance_matrix = distance.squareform(distances)  # Distance matrix.

# Measure the average and standard
# deviation of the nearest neighbor
# distances and find the nearest neighbor:

neighbor_distances = []
for i, row in enumerate(distance_matrix):
    nearest_dist = np.partition(row[row > 0], 0)[0]
    neighbor_distances.append(nearest_dist)

average_distance_pixels = np.mean(neighbor_distances)
std_dev_distance_pixels = np.std(neighbor_distances)

# Conversion factor from pixels to micrometers.
# Approximately 0.1111 micrometers per pixel:

micrometers_per_pixel = 20 / 180

# Convert all measurements to micrometers:

average_island_size_um = average_island_size * (micrometers_per_pixel ** 2)
min_island_size_um = min_island_size * (micrometers_per_pixel ** 2)
average_diameter_um = average_diameter_pixels * micrometers_per_pixel
average_distance_um = average_distance_pixels * micrometers_per_pixel
std_dev_distance_um = std_dev_distance_pixels * micrometers_per_pixel

# Identify and draw shapes around
# islands larger than 1.5 times
# the average diameter:

threshold_diameter = 1.5 * average_diameter_pixels
highlight_image = image.copy()

for size, centroid, label_index in valid_islands:
    diameter = 2 * np.sqrt(size / np.pi)
    if diameter > threshold_diameter:
        # Draw a bounding box around the large island:
        x, y, w, h = cv2.boundingRect((labels == label_index).astype(np.uint8))
        cv2.rectangle(highlight_image, (x, y), (x + w, y + h), (0, 0, 0), 6)  # Black rectangle!


# Output results under the figure:

print(f"Number of valid white islands (nuclei of cells) in the image: {valid_island_count}")
print(f"Average island size: {average_island_size_um:.2f} micrometers^2")
print(f"Minimum island size considered (average size / 2): {min_island_size_um:.2f} micrometers^2")
print(f"Average island diameter: {average_diameter_um:.2f} micrometers")
print(f"Average distance between neighboring nuclei: {average_distance_um:.2f} micrometers")
print(f"Standard deviation of distances: {std_dev_distance_um:.2f} micrometers")
print(f"Threshold for drawing shapes (1.5x average diameter): {threshold_diameter *
micrometers_per_pixel:.2f} micrometers")

# Visualize the refined mask
# with highlighted large islands:

plt.figure(figsize=(8, 8))
plt.imshow(highlight_image)
plt.title(f'Refined Otsu Thresholding Mask - Valid Cell Count: {valid_island_count}')
plt.axis('on')
plt.show()
```

Python implementation 27: Detection of abnormally sized nuclei for malignant tumor identification.

Connected component analysis is then performed to identify distinct nuclei within the refined binary mask. The code calculates the size (area) of each detected nucleus and filters out any that are smaller than half of the average size. This filtering process reduces noise and artifacts, ensuring that the analysis focuses only on relevant cellular structures. The remaining "valid" nuclei are analyzed further to compute their diameters and the distances between neighboring nuclei. The average diameter and average distance provide insights into the typical size and spatial arrangement of nuclei within the tissue sample. A significant addition in this code is the identification of nuclei with diameters larger than 1.5 times the average diameter. These nuclei are considered abnormally large and are potentially indicative of malignancy (Fig. 6.5). Next, Python Implementation 6.7 draws bounding boxes around these large nuclei to visually highlight them in the output image (Fig. 6.5). This feature is particularly useful for pathologists and researchers, as it allows for rapid identification of abnormal regions that may require further investigation or diagnostic attention. Therefore, let us notice the measurement values in the console.

Output

```
Number of valid white islands (nuclei of cells) in the image: 208
Average island size: 26.94 micrometers^2
Minimum island size considered (average size / 2): 13.47 micrometers^2
Average island diameter: 8.51 micrometers
Average distance between neighboring nuclei: 11.87 micrometers
Standard deviation of distances: 3.34 micrometers
Threshold for drawing shapes (1.5x average diameter): 12.77 micrometers
```

The *number of valid white islands* (nuclei of cells) in the image is 208, indicating that after filtering out small artifacts and noise, there are 208 nuclei that are considered significant for analysis. This count provides a measure of cellular density within the tissue, which can be an indicator of cellular proliferation or other pathological changes. Next, the *average island size* is 26.94 μm^2, and the *minimum island size* considered is 13.47 μm^2. These values give a sense of the overall size distribution of the nuclei, with the minimum size threshold helping to exclude smaller, irrelevant regions from being considered as potential nuclei. The average size is larger than what might be expected in normal tissue, which could suggest an increased presence of larger, possibly abnormal nuclei. The *average island diameter* is 8.51 μm, and the *average distance between neighboring nuclei* is 11.87 μm. These measurements reflect the average size of the nuclei and their spatial arrangement within the tissue. In a healthy tissue sample, these values are expected to fall within a certain range, and significant deviations may indicate abnormal tissue architecture, as seen in tumor growth or other pathological conditions. The relatively larger average distance suggests that while some nuclei are larger, they are not necessarily tightly packed, which can be characteristic of certain types of tumors (Fig. 6.5).

The *standard deviation* of distances between neighboring nuclei is 3.34 μm, indicating moderate variability in the spacing between nuclei. A higher standard deviation might suggest a more heterogeneous tissue architecture, where cells are unevenly spaced. In

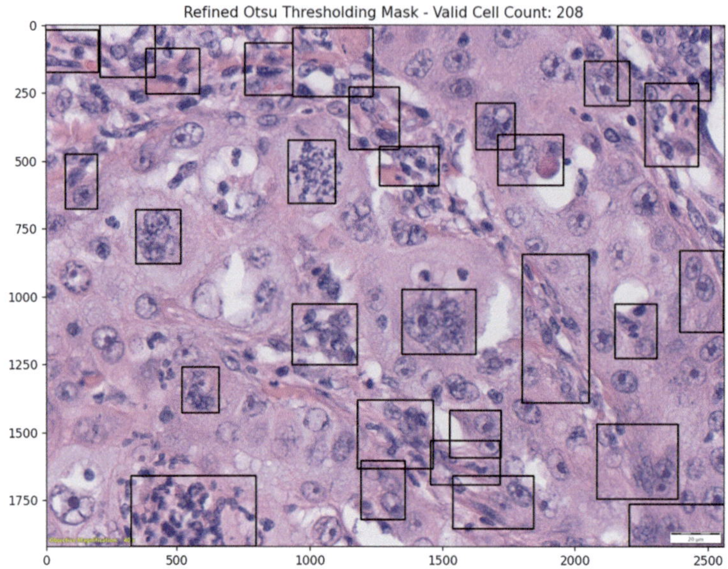

Fig. 6.5 Detection of abnormally sized nuclei for malignant tumor identification. The figure shows the detection of abnormally sized nuclei in a histological section of squamous cell carcinoma (SCC) in the skin of a cat. The original image has been processed using the "highlight large islands" technique to identify nuclei that are significantly larger than the average size. At 40× magnification with a scale bar of 20 μm, the figure demonstrates areas where nuclei exceed 1.5 times the average diameter, as indicated by black rectangles surrounding these regions. The analysis quantifies the variability in cell distribution, with a standard deviation of distances between neighboring nuclei at 3.34 μm, indicating moderate variability in spacing. This measure suggests a somewhat heterogeneous tissue architecture, where the cells are not uniformly distributed. A threshold for drawing shapes was set at 12.77 μm, representing 1.5 times the average diameter of the nuclei. Nuclei with diameters beyond this threshold are flagged as abnormally large. The highlighted nuclei in the final visual output help pinpoint areas of potential pathological concern where abnormal cell growth or tumor progression may be taking place. This approach allows for better visualization and quantification of potentially malignant cells in histological samples

contrast, a lower standard deviation might reflect a more uniform cell distribution. The *threshold for drawing shapes* is 12.77 μm, which represents 1.5 times the average diameter. Nuclei with diameters exceeding this threshold are considered abnormally large and are highlighted in the output image (Fig. 6.5). The highlighted nuclei, shown in the final visual output, draw attention to potential areas of concern where abnormal cell growth may be occurring (Fig. 6.5).

6.11 Detection of Malignant Nuclei by Filtering Inflammatory Cell Infiltrates

As discussed above, not all large nuclei are part of a possible tumor. There are also inflammatory and tissue infiltrates, for various reasons. The nucleus of these cells can be easily classified by our algorithm as part of a possible tumor. However, such leaks can be filtered out to make the final result closer to reality. Therefore, this next source code builds upon the previous example by not only detecting abnormally sized nuclei in histological images to identify potential malignant tumors but also by incorporating a new filtering step based on the roundness of these nuclei (Python Implementation 6.8). The focus is to eliminate inflammatory cell infiltrates from the results, as these infiltrates are often characterized by disordered and irregularly shaped cell clusters rather than round or oval nuclei. The initial steps of the code are similar to the previous approach. The histological image is loaded and converted to grayscale to simplify the subsequent processing steps. Then, *Otsu* thresholding is used here, again, to create a binary mask that separates the nuclei (white regions) from the background. Morphological operations are then used to refine this binary mask by removing small objects and filling in small holes, thus ensuring that only significant and well-defined nuclei are included in the analysis. Next, *connected component analysis* is performed to identify distinct nuclei in the refined binary mask. This step calculates the size (area) of each detected nucleus, filtering out those smaller than half of the average size, ensuring that the focus remains on significant cellular structures. This filtering step reduces noise and helps in excluding irrelevant small regions. The remaining acceptable nuclei are then analyzed to compute their diameters and the distances between neighboring nuclei, similar to the previous code (Python Implementation 6.7). These calculations provide insights into the size distribution and spatial arrangement of nuclei.

Enhanced detection of malignant nuclei

```
import cv2
import numpy as np
import matplotlib.pyplot as plt
from skimage.io import imread
from skimage.filters import threshold_otsu
from skimage.morphology import remove_small_objects, remove_small_holes, closing, square
from scipy.spatial import distance

image_path = r'C:\Users\Elitebook\Desktop\his_img.jpg'
image = imread(image_path)
gray_image = cv2.cvtColor(image, cv2.COLOR_RGB2GRAY)

otsu_thresh = threshold_otsu(gray_image)
mask_otsu = (gray_image < otsu_thresh).astype(np.uint8) * 255

mask_otsu_refined = remove_small_objects(mask_otsu.astype(bool), min_size=64).astype(np.uint8) * 255
mask_otsu_refined = remove_small_holes(mask_otsu_refined.astype(bool),
area_threshold=64).astype(np.uint8) * 255
mask_otsu_refined = closing(mask_otsu_refined, square(3))

num_labels, labels, stats, centroids = cv2.connectedComponentsWithStats(mask_otsu_refined,
connectivity=8)

island_sizes = stats[1:, cv2.CC_STAT_AREA]
average_island_size = np.mean(island_sizes)

min_island_size = average_island_size / 2

# Filter out islands smaller than the
# minimum size and keep their centroids:
valid_islands = [(size, centroid, i + 1) for i, (size, centroid) in enumerate(zip(island_sizes,
centroids[1:])) if size >= min_island_size]
valid_island_count = len(valid_islands)

# Calculate average diameter in pixels for valid islands:
diameters = [2 * np.sqrt(size / np.pi) for size, _, _ in valid_islands]
average_diameter_pixels = np.mean(diameters)

# Calculate distances between neighboring centroids:
centroids_filtered = np.array([centroid for _, centroid, _ in valid_islands])
distances = distance.pdist(centroids_filtered)    # Pairwise distances.
distance_matrix = distance.squareform(distances)  # Distance matrix.

# Measure the average and standard deviation
# of the nearest neighbor distances:
neighbor_distances = []
for i, row in enumerate(distance_matrix):
    nearest_dist = np.partition(row[row > 0], 0)[0]
    neighbor_distances.append(nearest_dist)

average_distance_pixels = np.mean(neighbor_distances)
std_dev_distance_pixels = np.std(neighbor_distances)

# Conversion - pixels to micrometers:
micrometers_per_pixel = 20 / 180

# Convert all measurements to micrometers:
average_island_size_um = average_island_size * (micrometers_per_pixel ** 2)
min_island_size_um = min_island_size * (micrometers_per_pixel ** 2)
average_diameter_um = average_diameter_pixels * micrometers_per_pixel
average_distance_um = average_distance_pixels * micrometers_per_pixel
std_dev_distance_um = std_dev_distance_pixels * micrometers_per_pixel

# Identify and draw shapes around islands
# larger than 1.5 times the average diameter
# and sufficiently round:
threshold_diameter = 1.5 * average_diameter_pixels
highlight_image = image.copy()

# Define circularity threshold:
circularity_threshold = 0.1

for size, centroid, label_index in valid_islands:
    diameter = 2 * np.sqrt(size / np.pi)
    if diameter > threshold_diameter:
        # Calculate perimeter to determine circularity:
        mask = (labels == label_index).astype(np.uint8)
        contours, _ = cv2.findContours(mask, cv2.RETR_EXTERNAL, cv2.CHAIN_APPROX_SIMPLE)
        if len(contours) > 0:
            perimeter = cv2.arcLength(contours[0], True)
            circularity = (4 * np.pi * size) / (perimeter ** 2) if perimeter != 0 else 0

            # Only draw red rectangle if circularity exceeds the threshold:
            if circularity >= circularity_threshold:
                x, y, w, h = cv2.boundingRect(contours[0])
                cv2.rectangle(highlight_image, (x, y), (x + w, y + h), (200, 0, 0), 6)

# Output results under the figure:
print(f"Number of valid white islands (nuclei of cells) in the image: {valid_island_count}")
print(f"Average island size: {average_island_size_um:.2f} micrometers^2")
print(f"Minimum island size considered (average size / 2): {min_island_size_um:.2f} micrometers^2")
print(f"Average island diameter: {average_diameter_um:.2f} micrometers")
print(f"Average distance between neighboring nuclei: {average_distance_um:.2f} micrometers")
print(f"Standard deviation of distances: {std_dev_distance_um:.2f} micrometers")
print(f"Threshold for drawing shapes (1.5x average diameter): {threshold_diameter *
micrometers_per_pixel:.2f} micrometers")
print(f"Circularity threshold for drawing: {circularity_threshold}")

# Visualize the refined
# mask with highlighted.
# large and sufficiently
# round islands:
plt.figure(figsize=(8, 8))
plt.imshow(highlight_image)
plt.title(f'Refined Otsu Thresholding Mask - Valid Cell Count: {valid_island_count}')
plt.axis('on')
plt.show()
```

Python implementation 28: Enhanced detection of malignant nuclei by using the roundness criteria.

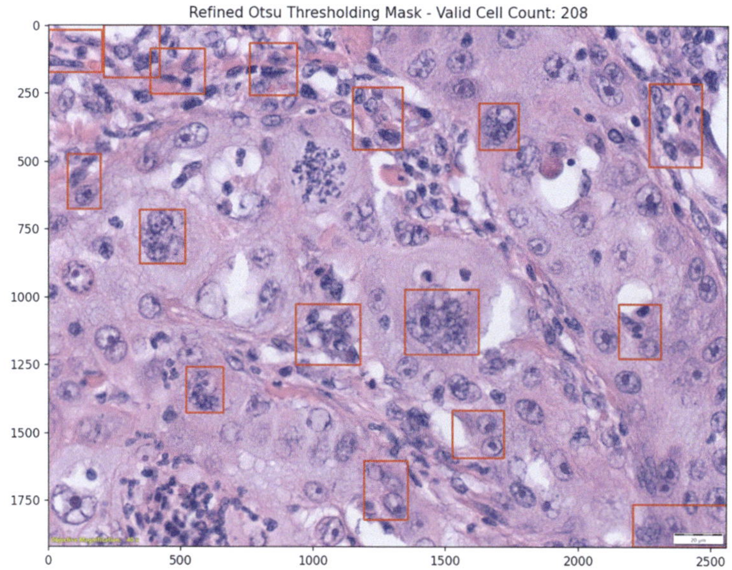

Fig. 6.6 Enhanced detection of malignant nuclei. The figure demonstrates the application of a roundness-based filtering technique to detect abnormally sized nuclei in a histological section of squamous cell carcinoma in a cat. At 40× magnification with a scale bar of 20 μm, the image highlights nuclei that exceed both size and circularity thresholds. The analysis measures an average island diameter of 8.51 μm and an average distance between neighboring nuclei of 11.87 μm. These values indicate that, while some nuclei are relatively large, they are not densely packed, which can be characteristic of certain tumor types. The standard deviation of distances between neighboring nuclei is 3.34 μm, suggesting moderate variability in the spacing between nuclei. A higher standard deviation might imply a more heterogeneous tissue architecture, while a lower standard deviation could reflect a more uniform cell distribution. The threshold for drawing shapes is set at 12.77 μm, which is 1.5 times the average diameter, indicating that nuclei with diameters exceeding this threshold are considered abnormally large. Additionally, a circularity threshold of 0.1 is applied to filter out non-oval or non-round shapes, such as those from inflammatory cell infiltrates, which tend to be disordered. Nuclei that meet both the size and circularity criteria are highlighted with red rectangles in the image. This combined size-based and shape-based filtering enhances the accuracy of detecting abnormal nuclei, providing a more precise and reliable tool for identifying potential malignancy in tissue samples. This approach may be particularly valuable for pathologists or researchers, as it focuses on identifying regions that may require further diagnostic investigation

The primary enhancement in this code (Python Implementation 6.8) involves filtering the nuclei based on their roundness or circularity. Roundness is an important parameter for distinguishing between normal and abnormal nuclei, as well as for differentiating malignant cells from inflammatory cell infiltrates (Fig. 6.6). Inflammatory cells often form disorganized clusters with irregular shapes that deviate significantly from the circular or oval shapes typical of nuclei. To filter out these inflammatory cell infiltrates, the code calculates the circularity of each nucleus that exceeds 1.5 times the average diameter. Circularity is calculated using the formula:

$$\text{Circularity} = \frac{4 \times \pi \times \text{Area}}{\text{Perimeter}^2}$$

A perfect circle has a circularity of 1.0, while shapes that are more elongated or irregular have values closer to 0. The code uses a circularity threshold (set at 0.1 in this example) to decide whether to highlight a nucleus. Only those nuclei that are both larger than 1.5 times the average diameter and have circularity above this threshold are considered abnormal and highlighted in the output image (Fig. 6.6). This refinement step is crucial for excluding inflammatory cell infiltrates, which do not meet the circularity criteria, thereby reducing false positives in the detection of abnormal nuclei indicative of malignancy. Next, the console output values can be more useful for conclusions.

Output

```
Number of valid white islands (nuclei of cells) in the image: 208
Average island size: 26.94 micrometers^2
Minimum island size considered (average size / 2): 13.47 micrometers^2
Average island diameter: 8.51 micrometers
Average distance between neighboring nuclei: 11.87 micrometers
Standard deviation of distances: 3.34 micrometers
Threshold for drawing shapes (1.5x average diameter): 12.77 micrometers
Circularity threshold for drawing: 0.1
```

Thus, the *number of valid white islands* (nuclei of cells) in the image is 208, just like before, indicating that after filtering out small artifacts and noise, there are 208 nuclei considered significant for analysis. This count gives an idea of the cellular density within the tissue, which can provide clues about cellular proliferation or other pathological changes. Next, the average island size is 26.94 μm^2, and the *minimum island size considered* is 13.47 μm^2. These values provide insight into the typical size of the nuclei within the tissue, with the minimum size threshold helping to exclude smaller, irrelevant regions from being considered as nuclei. The relatively large average size may suggest an increased presence of larger, potentially abnormal nuclei. The *average island diameter* is 8.51 μm, and the *average distance between neighboring nuclei* is the same as before, namely 11.87 μm. These measurements reflect the average size of the nuclei and their spatial arrangement within the tissue. The average distance suggests that while some nuclei are relatively large, they are not necessarily densely packed, which can be characteristic of certain types of tumors. It is worth mentioning again that the *standard deviation* of distances between neighboring nuclei is 3.34 μm, indicating moderate variability in the spacing between nuclei. Therefore, a higher standard deviation might suggest more heterogeneous tissue architecture, while a lower standard deviation might reflect a more uniform cell distribution.

The *threshold* for drawing shapes is 12.77 μm, representing 1.5 times the average diameter. Nuclei with diameters exceeding this threshold are considered abnormally large. Additionally, the *circularity threshold* for drawing is set at 0.1. Only those nuclei that meet both the size and circularity criteria are highlighted in the output image, as shown by the red rectangles in the final visualization. This approach allows for the detection of potential malignant cells while filtering out inflammatory cell infiltrates that do not meet the circularity criteria. Thus, this code enhances the accuracy of detecting abnormal nuclei in histological images by combining size-based and shape-based filtering, thus providing a more focused and reliable tool for identifying malignancy in tissue samples. This refinement may be particularly valuable for pathologists or researchers, as it allows for a more precise identification of abnormal regions that may require further diagnostic attention or investigation.

7 Machine Learning and AI in Microscopy Image Analysis

Abstract

Machine learning (ML) and artificial intelligence (AI) have revolutionized microscopy image analysis, enabling more accurate, efficient, and automated interpretations of complex biological data. This chapter shows practical examples in Python and discusses the application of ML, particularly deep learning, in veterinary pathology, with a focus on histopathological image analysis. Convolutional neural networks (CNNs), a subset of deep learning, have shown great promise in classifying tissue samples, detecting cellular structures, and identifying pathological changes without requiring manual feature extraction. Techniques such as U-Net facilitate precise image segmentation, crucial for quantifying cellular structures in dense tissues. Additionally, generative adversarial networks (GANs) are used to enhance image resolution, improving diagnostic accuracy. A key challenge remains in the availability of large, high-quality annotated datasets necessary for training ML models, though transfer learning offers a solution by enabling models pretrained on large datasets to be adapted for histopathological data. Despite their accuracy, deep learning models often operate as "black boxes," spurring the development of explainable AI techniques like Grad-CAM, which provide visual interpretations of model decisions. These advancements not only enhance diagnostic precision but also increase the transparency and trustworthiness of AI systems in veterinary diagnostics, marking a significant shift in the field.

Keywords

Machine learning · Neural networks · Convolutional neural networks · Deep learning · Transfer learning

E. Gagniuc, *Computational Analysis of Microscopy Images*,
https://doi.org/10.1007/978-3-031-86270-0_7

7.1 Introduction

Machine learning and artificial intelligence (AI) have revolutionized the field of microscopy image analysis by enabling more accurate, efficient, and automated interpretations of complex biological data [84]. The rapid advancements in computational power, combined with the availability of large-scale annotated datasets, have paved the way for applying sophisticated machine learning algorithms to analyze histopathological images [79]. These approaches have shown significant promise in detecting subtle patterns that may not be easily discernible through traditional methods, thereby enhancing the diagnostic process and supporting clinical decision-making in both human and veterinary medicine [102]. Deep learning, a subset of machine learning, has become a cornerstone in the analysis of microscopy images. *Convolutional neural networks* (CNNs), for instance, are specifically designed to handle image data by learning spatial hierarchies of features through multiple layers of convolutional filters [102]. This architecture makes CNNs particularly effective for tasks such as image classification, segmentation, and object detection. In the context of histology, CNNs can be trained to classify tissue samples, detect cellular structures, and identify pathological changes [80]. When applied to the detection of malignant tumors, CNNs can differentiate between normal and abnormal cell morphologies based on features learned directly from the data, without requiring manual feature extraction [10, 82, 84]. This ability to learn directly from raw data represents a significant shift from traditional image analysis techniques, where handcrafted features were used to define morphological characteristics. Another powerful application of machine learning in microscopy is image segmentation. Accurate segmentation is crucial for quantifying cellular and subcellular structures in histological images. *U-Net*, a type of neural network specifically designed for biomedical image segmentation, has become a popular choice in this domain [152]. The *U-Net* architecture enables precise localization and classification of structures within an image by using a contracting path to capture context and a symmetric expanding path to enable precise localization. This design is particularly effective in separating overlapping cells or structures that are densely packed, a common challenge in histological analysis [16]. Machine learning algorithms also play a critical role in enhancing image quality and resolution. Techniques such as generative adversarial networks (GANs) have been employed to perform super-resolution imaging, where low-resolution images are converted into high-resolution images that reveal fine structural details [153]. This capability is particularly useful when the original microscopy images are of suboptimal quality due to technical limitations or artifacts. Image quality improvement allows GANs and other super-resolution techniques to enable more accurate segmentation, feature extraction, and analysis, ultimately leading to better diagnostic performance. The addition of machine learning into microscopy image analysis also allows for the development of predictive models that can assist in prognostication and treatment planning. For example, machine learning models can be trained to predict patient outcomes based on histological features extracted from tissue samples. One of the key challenges in applying machine learning to microscopy image analysis is the need for large, high-quality annotated datasets. The performance of machine learning models, particularly deep learning models, is heavily dependent on the availability of labeled data for training.

In histopathology, generating such datasets is time-consuming and requires expert pathologist input to ensure accurate annotations. Efforts to create large-scale public datasets, combined with advances in transfer learning, where models pretrained on large generic datasets are fine-tuned on specific histopathological data, have been instrumental in overcoming this limitation. Transfer learning enables the adaptation of existing models to new datasets with limited labeled data, significantly reducing the time and effort required for training while maintaining high performance. Another consideration is the interpretability of machine learning models. While deep learning models, particularly CNNs, have shown exceptional accuracy in image analysis tasks, they are often criticized as "black boxes" due to their complex internal structures that are not easily interpretable. This lack of transparency can be a barrier to clinical adoption, where understanding the rationale behind a model decision is crucial. Recent advances in explainable AI (XAI) seek to address this issue by developing methods to interpret and visualize the decision-making process of machine learning models. Techniques such as *Grad-CAM* (gradient-weighted class activation mapping) provide visual explanations by highlighting the regions of an image that contribute most to the prediction of the model, thereby making the models more reliable. Worthy of note is that AI technologies not only facilitate advanced image analysis but also integrate genetic and biochemical data, such as in studies of diabetes, to provide a comprehensive diagnostic and prognostic approach in veterinary medicine [154–156].

7.2 Introduction to Machine Learning in Image Analysis

Machine learning has emerged as a powerful tool in image analysis, transforming how we interpret and utilize visual data across various fields, including histopathology and microscopy [79]. In the context of image analysis, machine learning involves developing algorithms that can automatically learn from and make decisions based on data. This capability is particularly useful in analyzing complex and high-dimensional data, such as images obtained from microscopes, where traditional rule-based methods fall short due to the variability and complexity inherent in biological samples. In histopathology, machine learning facilitates the automated extraction of meaningful patterns and features from images, enabling more accurate, efficient, and reproducible analysis. The most common type of machine learning used in image analysis is supervised learning, where algorithms are trained on labeled datasets to recognize specific patterns, such as cellular structures, tissue types, or pathological markers. One of the most impactful subsets of machine learning in image analysis is deep learning, which uses neural networks with multiple layers to learn hierarchical representations of data. *Convolutional neural networks* (CNNs), a class of deep learning models, have become the standard approach for image analysis due to their ability to automatically learn spatial hierarchies of features, from low-level features like edges and textures to high-level features such as shapes and patterns. CNNs require minimal preprocessing compared to traditional methods and can handle large amounts of data, making them particularly well suited for processing high-resolution microscopy images. Machine learning also enables advanced image segmentation, which is the process of partitioning an image into meaningful regions, such as separating nuclei from the background or distinguishing between different types of tissues. Traditional segmentation techniques, such as thresholding or edge detection, often struggle with variability in staining, illumination, and tissue morphology. Machine learning models, however, can be trained on various datasets to learn how to adapt to these variations, providing more robust and accurate segmentation outcomes. This capability is crucial in histopathology, where consistent segmentation is foundational for downstream analysis, such as feature extraction, cell counting, and morphology assessment. In addition to segmentation, machine learning algorithms excel at image classification, which involves categorizing entire images or specific regions within images based on their visual characteristics. This has direct applications in diagnostic pathology, where images need to be classified into categories such as benign, precancerous, or malignant. Another significant application of machine learning in image analysis is in the quantification of features, where algorithms extract numerical data from images to provide insights into biological processes. For instance, machine learning can be used to quantify the density of immune cell infiltration in a tumor microenvironment or measure the heterogeneity of nuclear size and shape in cancerous tissues. These quantitative metrics can be invaluable for assessing disease progression, evaluating treatment responses, and identifying prognostic or predictive biomarkers. Machine learning also plays a critical role in enhancing image quality through techniques such as image restoration and denoising. In microscopy, images often suffer from noise due to limitations in imaging hardware or conditions. Machine learning models, particularly those using generative adversarial networks (GANs) or autoencoders, can learn to reconstruct high-quality images from noisy inputs, significantly improving the clarity and usability of microscopy data for analysis [157]. Despite its transformative potential, implementing machine learning in image analysis comes with challenges. The need for large, annotated datasets to train robust models is a significant bottleneck. Annotating such data requires expert input and is time-consuming and labor-intensive. Additionally, ensuring the generalizability of machine learning models across different datasets and imaging platforms remains a key concern, necessitating rigorous validation and possibly domain adaptation techniques. The interpretability of machine learning models, particularly deep learning, is another critical issue. While these models can achieve high accuracy, understanding how they arrive at their decisions is often not straightforward. Developing more interpretable models or integrating explainability methods is essential for clinical adoption, where transparency and trust in automated decision-making systems are paramount.

7.3 Convolutional Neural Networks (CNNs) for Image Classification

Convolutional neural networks (CNNs) have emerged as a fundamental tool for image classification tasks in microscopy, particularly for analyzing histopathological images. CNNs are a type of deep learning model specifically designed to work with grid-like data, such as images, by leveraging spatial hierarchies and local connectivity patterns. In the context of image classification, CNNs can learn complex patterns and features from raw pixel data to differentiate between various categories, such as healthy and diseased tissue, different types of cells, or distinct subcellular structures. The architecture of a CNN is inspired by the organization of the visual cortex in the human brain, where neurons respond to stimuli in a localized receptive field. This biological inspiration allows CNNs to effectively capture spatial hierarchies of features in images. The typical structure of a CNN consists of multiple layers, including convolutional layers, pooling layers, and fully connected layers. Convolutional layers are responsible for automatically extracting features from the input image using learnable filters, which slide across the input to produce feature maps. These filters are trained to detect low-level features, such as edges and textures, in the initial layers, while deeper layers capture more complex patterns, such as shapes and objects. Pooling layers, often interspersed between convolutional layers, are used to down-sample the feature maps, reducing the dimensionality and computational complexity while preserving the most salient information. This process, known as spatial down-sampling, helps to make the network more robust to small translations and distortions in the input image. Max pooling, which takes the maximum value from a region of the feature map, is one of the most commonly used pooling strategies in CNNs. After a series of convolutional and pooling layers, the high-level abstract features are passed through fully connected layers that perform the actual classification by mapping the extracted features to class probabilities. In microscopy image analysis, CNNs are used to classify various histopathological features. For example, they can distinguish between different types of cells (such as epithelial, lymphocytes, and stromal cells), identify cancerous versus noncancerous regions, and categorize tumor subtypes based on morphological patterns. One of the key advantages of CNNs is their ability to learn directly from raw image data, eliminating the need for manual feature engineering. This capability is particularly valuable in histology, where defining specific features for every possible tissue or cellular pattern is challenging. Instead, CNNs learn discriminative features automatically through backpropagation, where the model adjusts the filter weights based on the error between the predicted and true labels during training. The performance of CNNs in image classification tasks can be significantly enhanced with data augmentation techniques. In histopathology, data augmentation involves artificially expanding the training dataset by applying random transformations to the input images, such as rotations, flips, zooms, and brightness adjustments [114]. This process helps prevent overfitting, where the model becomes too specialized to the training data and performs poorly on new, unseen data. *Transfer learning* is another powerful strategy that can be used to improve the performance of CNNs in microscopy image classification. Transfer learning leverages pretrained CNN models, such as *VGG*, *ResNet*, and *Inception*, which have been trained on large-scale image datasets like *ImageNet* [158]. These models already contain filters that capture generic image features, such as edges and textures, which can be relevant across different domains. One of the main challenges in applying CNNs to microscopy images is the variability and heterogeneity of histological data. Different staining techniques, magnification levels, and sample preparations can introduce significant variability, making it difficult for CNNs to generalize well. To address this issue, CNN-based models are often trained using many different datasets that capture a wide range of histological appearances, thereby enhancing their robustness to variations. Moreover, the use of stain normalization techniques, which standardize the color distribution of histological images, can help reduce variability and improve the consistency of CNN performance across different datasets. Explainability and interpretability are also important considerations when applying CNNs for image classification in microscopy. While CNNs have shown high accuracy in detecting and classifying histopathological features, their "black-box" nature poses a challenge in clinical settings, where understanding the rationale behind the prediction of a model is crucial for gaining trust from clinicians and pathologists. Techniques such as *class activation mapping* (CAM) and *gradient-weighted class activation mapping* (Grad-CAM) provide visual explanations by highlighting the regions of an image that contribute most to a CNN decision. Thus, these interpretability methods help bridge the gap between the high accuracy of CNN models and the need for transparency in medical diagnostics.

7.3.1 Training a Deep Learning Model for Automated Detection of Tumors

Computer-aided diagnosis in veterinary pathology using a *convolutional neural network* (CNN) is one of the advanced methods of today that veterinary medicine is slowly adapting to. Therefore, the new example found in Python Implementation 7.1 is designed for training a CNN to classify histological images as either "normal" or "tumor" based on a dataset of cat skin tissue, specifically localized to the ear with sarcomas of soft tissue, imaged at 40× magnification. The images used for training were carefully chosen: 15 images of normal tissue and 21 images of tumoral tissue, each containing a scale bar of 20 μm for accurate scaling and interpretation. Before running this code, the reader must set up the correct environment. This includes having Python installed along with the necessary libraries such as *TensorFlow*, *NumPy*, *tkinter*, and *matplotlib*. The reader should also ensure the training and validation directories (train and validation) are properly set up, containing subdirectories named "normal" and "tumoral," each filled with the corresponding images for training and validation. The implementation is structured into several steps that take the example from data preparation, model building, training, and validation to making predictions on new images. Python Implementation 7.1 starts with setting up the data generators using *ImageDataGenerator* from *tensorflow.keras*, which is responsible for real-time data augmentation such as rescaling, rotation, width and height shifting, shearing, zooming, and horizontal flipping. These augmentations improve the generalizability of the model by preventing overfitting on the limited dataset of 36 images. The CNN model built in the code is a sequential model comprising three convolutional layers with increasing filter sizes (32, 64, 128), each followed by a max-pooling layer to downsample the feature maps. After these layers, the feature maps are flattened, followed by a dense layer with 128 units and a dropout layer with a rate of 0.5 to further prevent overfitting. The output layer is a single neuron with a sigmoid activation function, suitable for binary classification tasks like this one.

CNN for classification of normal and tumor tissue images

```
import os
import numpy as np
import matplotlib.pyplot as plt
from tensorflow.keras.preprocessing.image import ImageDataGenerator
from tensorflow.keras.models import Sequential, load_model
from tensorflow.keras.layers import Conv2D, MaxPooling2D, Flatten, Dense, Dropout
from tensorflow.keras.callbacks import EarlyStopping
from tensorflow.keras.preprocessing import image
import tkinter as tk
from tkinter import filedialog

# Step 1: Set up directories for
# training and validation data:

train_dir = 'C:\\Users\\Elitebook\\Desktop\\train'
val_dir = 'C:\\Users\\Elitebook\\Desktop\\validation'

# Step 2: Data generators for
# training and validation:

train_datagen = ImageDataGenerator(
    rescale=1.0/255.0,
    rotation_range=20,
    width_shift_range=0.2,
    height_shift_range=0.2,
    shear_range=0.2,
    zoom_range=0.2,
    horizontal_flip=True
)

val_datagen = ImageDataGenerator(rescale=1.0/255.0)

train_generator = train_datagen.flow_from_directory(
    train_dir,
    target_size=(128, 128),
    batch_size=8,
    class_mode='binary'
)

val_generator = val_datagen.flow_from_directory(
    val_dir,
    target_size=(128, 128),
    batch_size=8,
    class_mode='binary'
)

# Step 3: Build the CNN model:
model = Sequential([
    Conv2D(32, (3, 3), activation='relu', input_shape=(128, 128, 3)),
    MaxPooling2D((2, 2)),
    Conv2D(64, (3, 3), activation='relu'),
    MaxPooling2D((2, 2)),
    Conv2D(128, (3, 3), activation='relu'),
    MaxPooling2D((2, 2)),
    Flatten(),
    Dense(128, activation='relu'),
    Dropout(0.5),
    Dense(1, activation='sigmoid')
])

model.compile(optimizer='adam', loss='binary_crossentropy', metrics=['accuracy'])
model.summary()

# Step 4: Train the model
# with early stopping:
early_stopping = EarlyStopping(monitor='val_loss', patience=5, restore_best_weights=True)

history = model.fit(
    train_generator,
    steps_per_epoch=len(train_generator),
    epochs=30,
    validation_data=val_generator,
    validation_steps=len(val_generator),
    callbacks=[early_stopping]
)

# Step 5: Save the trained model:
model_save_path = 'C:\\Users\\Elitebook\\Desktop\\cat_histology_model.h5'
model.save(model_save_path)
print(f"Model saved to {model_save_path}")

# Step 6: Function to load the
# model and predict on new images:
def predict_image(model_path):
    # Load the model
    model = load_model(model_path)

    # Create a file dialog to select an image:
    root = tk.Tk()
    root.withdraw()  # Hide the root window.
    file_path = filedialog.askopenfilename(title="Select an Image for Prediction",
                                           filetypes=[("Image files", "*.jpg *.jpeg *.png")])

    if not file_path:
        print("No image selected. Exiting...")
        return

    # Load and preprocess the selected image:
    img = image.load_img(file_path, target_size=(128, 128))
    img_array = image.img_to_array(img)
    # Expand dimensions:
    img_array = np.expand_dims(img_array, axis=0)
    img_array /= 255.0  # Normalize.

    # Predict using the model:
    prediction = model.predict(img_array)

    # Output the result:
    if prediction[0] > 0.5:
        print(f"The model predicts the image is **Tumor** with " \
              f"a confidence of {prediction[0][0] * 100:.2f}%")
    else:
        print(f"The model predicts the image is **Normal** with" \
              f"a confidence of {(1 - prediction[0][0]) * 100:.2f}%")

# Step 7: Run the
# prediction function:
predict_image(model_save_path)
```

Python implementation 29: A *CNN* for classification of normal and tumor tissue images.

The model is compiled with the *Adam* optimizer, *binary cross-entropy loss* function, and accuracy as a metric. Note that the *Adam* optimizer is an adaptive learning rate optimization algorithm used in training deep learning models [159]. It combines the benefits of two other popular methods, *AdaGrad* and *RMSProp*, to provide fast convergence and efficient computation by adjusting the learning rate for each parameter dynamically. This makes it well suited for large datasets and high-dimensional parameter spaces. The *binary cross-entropy loss* function, also known as log loss, is used for binary classification problems [160]. It measures the difference between the predicted probabilities and the actual binary labels (0 or 1). The function penalizes incorrect predictions more heavily, making it ideal for training models where the goal is to output probabilities for two possible outcomes. The last notion, namely the *accuracy as a metric* evaluates the performance of a model by calculating the proportion of correct predictions out of the total number of predictions [161]. It is simple to interpret and provides a straightforward indication of how well the model is performing in terms of correctly classifying data points, particularly useful when the data is balanced between classes. Next, in Python Implementation 7.1, training is done with early stopping, a technique that monitors the validation loss and stops training when the loss stops decreasing for five consecutive epochs, ensuring the best weights are restored. This technique is essential when working with limited datasets to avoid overfitting. After training for up to 30 epochs, the model is saved to disk. The saved model can then be used to make predictions on new histological images through a graphical user interface (GUI)-based file dialog created using *tkinter*. The *predict_image* function loads the pretrained model and allows the user to select an image for prediction. The selected image is preprocessed (resized to 128 × 128, normalized) and fed into the model for prediction. The output is a classification of the image as "tumor" or "normal" along with the confidence percentage. Python Implementation 7.1 is valuable in a veterinary pathology setting, where quick and accurate diagnostics of histological samples can aid in early detection and treatment planning for conditions like sarcomas in cats. The use of 40× magnification with a 20-μm scale bar ensures that the model is trained on high-resolution images, capturing critical cellular and tissue-level features (Figs. 7.1 and 7.2). These features are vital for distinguishing between normal and tumorous tissues, especially when differentiating between malignant and benign growths. The reader must ensure that the image data is carefully annotated and prepared, as this affects the performance of the model significantly. Understanding the magnification and scale bar implications helps in interpreting the focus and resolution of the model, ensuring that it generalizes well to new samples. Furthermore, it is important to consider that this model is highly specific to the trained dataset and may not generalize to other types of tissue or different magnifications without retraining or transfer learning (please see the next section). Please note that the console output provides a detailed log of the model training and prediction process for classifying histological images (of cat ear tissue in this case) into "normal" or "tumor." Here, the prediction process is only for testing purposes because later in this chapter the training is used for diagnosis (Python Implementation 7.2).

Output

```
PS C:\Users\Elitebook>
& C:/Users/Elitebook/AppData/Local/Programs/Python/Python312/python.exe "c:/Users/Elitebook/Desktop/1 generate train file and
recognize.py"
2024-09-11 20:48:54.136913: I tensorflow/core/util/port.cc:153] oneDNN custom operations are on. You may see slightly different
numerical results due to floating-point round-off errors from different computation orders. To turn them off, set the environment
variable `TF_ENABLE_ONEDNN_OPTS=0`.
2024-09-11 20:48:55.792259: I tensorflow/core/util/port.cc:153] oneDNN custom operations are on. You may see slightly different
numerical results due to floating-point round-off errors from different computation orders. To turn them off, set the environment
variable `TF_ENABLE_ONEDNN_OPTS=0`.
Found 36 images belonging to 2 classes.
Found 2 images belonging to 2 classes.
C:\Users\Elitebook\AppData\Local\Programs\Python\Python312\Lib\site-packages\keras\src\layers\convolutional\base_conv.py:107:
UserWarning: Do not pass an `input_shape`/`input_dim` argument to a layer. When using Sequential models, prefer using an
`Input(shape)` object as the first layer in the model instead.
super().__init__(activity_regularizer=activity_regularizer, **kwargs)
2024-09-11 20:49:00.222340: I tensorflow/core/platform/cpu_feature_guard.cc:210] This TensorFlow binary is optimized to use available
CPU instructions in performance-critical operations.
To enable the following instructions: AVX2 FMA, in other operations, rebuild TensorFlow with the appropriate compiler flags.
Model: "sequential"
```

Layer (type)	Output Shape	Param #
conv2d (Conv2D)	(None, 126, 126, 32)	896
max_pooling2d (MaxPooling2D)	(None, 63, 63, 32)	0
conv2d_1 (Conv2D)	(None, 61, 61, 64)	18,496
max_pooling2d_1 (MaxPooling2D)	(None, 30, 30, 64)	0
conv2d_2 (Conv2D)	(None, 28, 28, 128)	73,856
max_pooling2d_2 (MaxPooling2D)	(None, 14, 14, 128)	0
flatten (Flatten)	(None, 25088)	0
dense (Dense)	(None, 128)	3,211,392
dropout (Dropout)	(None, 128)	0
dense_1 (Dense)	(None, 1)	129

```
Total params: 3,304,769 (12.61 MB)
Trainable params: 3,304,769 (12.61 MB)
Non-trainable params: 0 (0.00 B)
Epoch 1/30
C:\Users\Elitebook\AppData\Local\Programs\Python\Python312\Lib\site-
packages\keras\src\trainers\data_adapters\py_dataset_adapter.py:122: UserWarning: Your `PyDataset` class should call
`super().__init__(**kwargs)` in its constructor. `**kwargs` can include `workers`, `use_multiprocessing`, `max_queue_size`. Do not
pass these arguments to `fit()`, as they will be ignored.
self._warn_if_super_not_called()
5/5 ──────────────────── 11s 363ms/step - accuracy: 0.3469 - loss: 0.7437 - val_accuracy: 0.5000 - val_loss: 0.7848
Epoch 2/30
2024-09-11 20:49:14.723279: I tensorflow/core/framework/local_rendezvous.cc:404] Local rendezvous is aborting with status:
OUT_OF_RANGE: End of sequence
[[{{node IteratorGetNext}}]]
C:\Users\Elitebook\AppData\Local\Programs\Python\Python312\Lib\contextlib.py:158: UserWarning: Your input ran out of data;
interrupting training. Make sure that your dataset or generator can generate at least `steps_per_epoch * epochs` batches. You may need
to use the `.repeat()` function when building your dataset.
self.gen.throw(value)
2024-09-11 20:49:14.738042: I tensorflow/core/framework/local_rendezvous.cc:404] Local rendezvous is aborting with status:
OUT_OF_RANGE: End of sequence
[[{{node IteratorGetNext}}]]
C:\Users\Elitebook\AppData\Local\Programs\Python\Python312\Lib\site-packages\keras\src\callbacks\early_stopping.py:155: UserWarning:
Early stopping conditioned on metric `val_loss` which is not available. Available metrics are: accuracy,loss
current = self.get_monitor_value(logs)
5/5 ──────────────────── 0s 9ms/step - accuracy: 0.0000e+00 - loss: 0.0000e+00
Epoch 3/30
5/5 ──────────────────── 8s 232ms/step - accuracy: 0.6233 - loss: 0.5882 - val_accuracy: 0.5000 - val_loss: 0.6650
Epoch 4/30
2024-09-11 20:49:23.155200: I tensorflow/core/framework/local_rendezvous.cc:404] Local rendezvous is aborting with status:
OUT_OF_RANGE: End of sequence
[[{{node IteratorGetNext}}]]
5/5 ──────────────────── 0s 2ms/step - accuracy: 0.0000e+00 - loss: 0.0000e+00
Epoch 5/30
5/5 ──────────────────── 8s 215ms/step - accuracy: 0.7355 - loss: 0.6621 - val_accuracy: 1.0000 - val_loss: 0.6130
Epoch 6/30
5/5 ──────────────────── 0s 3ms/step - accuracy: 0.0000e+00 - loss: 0.0000e+00
Epoch 7/30
5/5 ──────────────────── 8s 216ms/step - accuracy: 0.6687 - loss: 0.6099 - val_accuracy: 1.0000 - val_loss: 0.4643
Epoch 8/30
2024-09-11 20:49:39.191422: I tensorflow/core/framework/local_rendezvous.cc:404] Local rendezvous is aborting with status:
OUT_OF_RANGE: End of sequence
[[{{node IteratorGetNext}}]]
5/5 ──────────────────── 0s 3ms/step - accuracy: 0.0000e+00 - loss: 0.0000e+00
Epoch 9/30
5/5 ──────────────────── 8s 217ms/step - accuracy: 0.7870 - loss: 0.5328 - val_accuracy: 1.0000 - val_loss: 0.3705
Epoch 10/30
5/5 ──────────────────── 0s 0s/step - accuracy: 0.0000e+00 - loss: 0.0000e+00
Epoch 11/30
5/5 ──────────────────── 8s 219ms/step - accuracy: 0.8128 - loss: 0.4282 - val_accuracy: 0.5000 - val_loss: 0.8619
Epoch 12/30
5/5 ──────────────────── 0s 0s/step - accuracy: 0.0000e+00 - loss: 0.0000e+00
Epoch 13/30
5/5 ──────────────────── 9s 225ms/step - accuracy: 0.7810 - loss: 0.4701 - val_accuracy: 1.0000 - val_loss: 0.2319
Epoch 14/30
5/5 ──────────────────── 0s 7ms/step - accuracy: 0.0000e+00 - loss: 0.0000e+00
Epoch 15/30
5/5 ──────────────────── 8s 234ms/step - accuracy: 1.0000 - loss: 0.3061 - val_accuracy: 1.0000 - val_loss: 0.0849
Epoch 16/30
2024-09-11 20:50:12.279352: I tensorflow/core/framework/local_rendezvous.cc:404] Local rendezvous is aborting with status:
OUT_OF_RANGE: End of sequence
[[{{node IteratorGetNext}}]]
5/5 ──────────────────── 0s 5ms/step - accuracy: 0.0000e+00 - loss: 0.0000e+00
Epoch 17/30
5/5 ──────────────────── 8s 213ms/step - accuracy: 1.0000 - loss: 0.1479 - val_accuracy: 1.0000 - val_loss: 0.1130
Epoch 18/30
5/5 ──────────────────── 0s 0s/step - accuracy: 0.0000e+00 - loss: 0.0000e+00
Epoch 19/30
5/5 ──────────────────── 8s 213ms/step - accuracy: 0.9487 - loss: 0.1621 - val_accuracy: 1.0000 - val_loss: 0.0066
Epoch 20/30
5/5 ──────────────────── 0s 3ms/step - accuracy: 0.0000e+00 - loss: 0.0000e+00
Epoch 21/30
5/5 ──────────────────── 9s 231ms/step - accuracy: 0.8877 - loss: 0.2838 - val_accuracy: 0.5000 - val_loss: 0.3943
Epoch 22/30
5/5 ──────────────────── 0s 2ms/step - accuracy: 0.0000e+00 - loss: 0.0000e+00
Epoch 23/30
5/5 ──────────────────── 8s 213ms/step - accuracy: 0.8398 - loss: 0.2299 - val_accuracy: 1.0000 - val_loss: 0.0090
Epoch 24/30
5/5 ──────────────────── 0s 0s/step - accuracy: 0.0000e+00 - loss: 0.0000e+00
Epoch 25/30
5/5 ──────────────────── 8s 242ms/step - accuracy: 0.9682 - loss: 0.1157 - val_accuracy: 1.0000 - val_loss: 0.0161
Epoch 26/30
5/5 ──────────────────── 0s 0s/step - accuracy: 0.0000e+00 - loss: 0.0000e+00
Epoch 27/30
5/5 ──────────────────── 8s 218ms/step - accuracy: 0.9348 - loss: 0.1532 - val_accuracy: 1.0000 - val_loss: 0.0064
Epoch 28/30
5/5 ──────────────────── 0s 2ms/step - accuracy: 0.0000e+00 - loss: 0.0000e+00
Epoch 29/30
5/5 ──────────────────── 8s 214ms/step - accuracy: 0.9239 - loss: 0.2847 - val_accuracy: 1.0000 - val_loss: 0.0349
Epoch 30/30
5/5 ──────────────────── 0s 0s/step - accuracy: 0.0000e+00 - loss: 0.0000e+00
WARNING:absl:You are saving your model as an HDF5 file via `model.save()` or `keras.saving.save_model(model)`. This file format is
considered legacy. We recommend using instead the native Keras format, e.g. `model.save('my_model.keras')` or
`keras.saving.save_model(model, 'my_model.keras')`.
Model saved to C:\Users\Elitebook\Desktop\cat_histology_model.h5
WARNING:absl:Compiled the loaded model, but the compiled metrics have yet to be built. `model.compile_metrics` will be empty until you
train or evaluate the model.
1/1 ──────────────────── 0s 90ms/step
The model predicts the image is **Tumor** with a confidence of 99.01%
PS C:\Users\Elitebook>
```

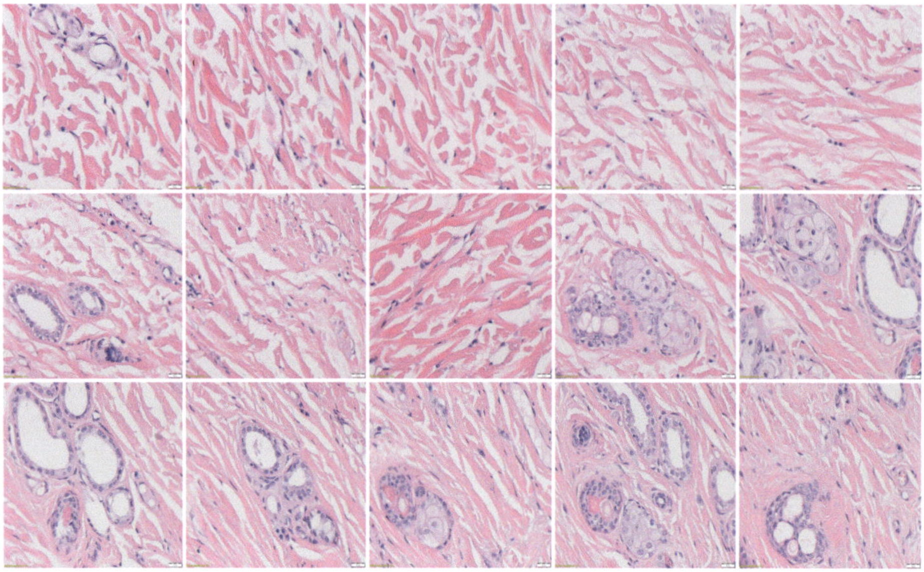

Fig. 7.1 Training images for the deep learning script, depicting histological sections of cat skin. These images represent normal tissue captured under 40× magnification, with a 20-μm scale bar shown in each image for reference. These histological images show the characteristic appearance of normal cat skin tissue, highlighting the extracellular matrix and distribution of nuclei. Several images also feature hair follicles as well as secretory glands, which are identifiable by their circular structures with dense cellular regions. The pink hue is due to standard hematoxylin and eosin (H&E) staining, where nuclei appear blue-purple, and the surrounding collagen fibers in the connective tissue are pink. This set of images is used for training purposes in machine learning models to distinguish normal skin from pathological states. The images are used to train a convolutional neural network (CNN) model for distinguishing between normal and tumoral tissues in cat ear histology

The first few lines indicate that *TensorFlow* uses *oneAPI Deep Neural Network Library* (oneDNN) custom operations for optimized performance on the CPU [162]. This information is typical when *TensorFlow* is setting up the environment and is generally not a concern for the user unless there are performance issues. If required, oneDNN optimizations can be turned off by setting the environment variable *TF_ENABLE_ONEDNN_OPTS = 0*. As a side note, *oneDNN* is an open-source performance library developed by Intel that provides optimized building blocks for deep learning applications. It supports a wide range of operations used in neural network training and inference, such as convolutions, pooling, normalization, and more. Also, *oneDNN* is designed to accelerate deep learning workloads on Intel CPUs and GPUs by optimizing low-level mathematical operations, ensuring efficient computation and memory usage. It is integrated into popular deep learning frameworks like *TensorFlow* and *PyTorch*, where it helps improve the performance of deep learning models on Intel hardware. Next, the console then confirms the loading of 36 images for training and 2 images for validation. These images are categorized into two classes, "normal" and "tumor" (Figs. 7.1 and 7.2). The validation set is very small (only

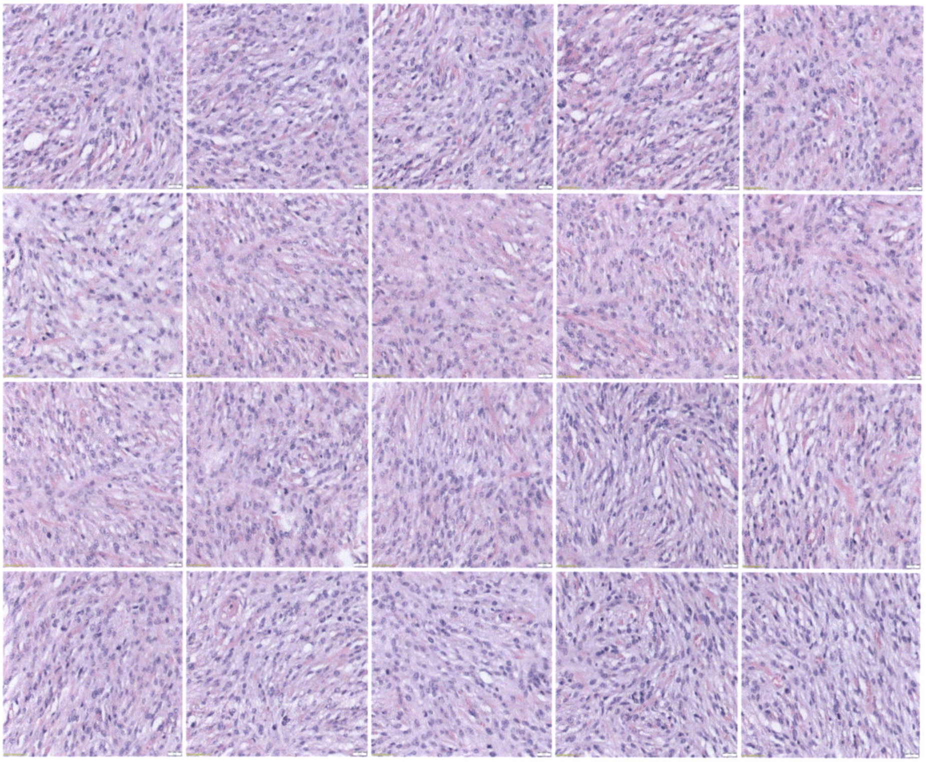

Fig. 7.2 Training images for the deep learning script, showing histological sections of cat skin infiltrated with tumoral tissue. These images represent abnormal, tumoral tissue captured under 40× magnification, with a 20-μm scale bar included in each image for size reference. The samples display dense, disordered cellular architecture typical of soft tissue sarcoma, demonstrating the presence of a tumor infiltrating the skin. These images are used to train a convolutional neural network (CNN) model for the differentiation of normal versus tumoral tissue in cat ear histology

two images), which might affect the performance of the model. However, for the exemplification here, these two images are sufficient, to avoid overcrowding the pages of this book. Also, a *UserWarning* appears next, suggesting that the input shape should be provided using an Input(shape) object rather than the *input_shape* argument in the sequential model.

This is a minor warning that does not affect the functionality of the model but could be addressed for better practice. In Keras, an *input (shape)* object is used to explicitly define the input layer for a neural network model [163]. It is a more flexible and recommended approach, particularly when building complex models or using the *Functional API*, as it clearly specifies the shape of the input data. The *input_shape* argument, on the other hand, is used directly in the first layer of a sequential model to define the input size. The next lines in the console provide the architecture of the CNN model, showing each type of layer, output shape, and number of parameters (weights and biases). The network consists of three convolutional layers followed by max pooling, a flattening layer, a dense (fully connected) layer with 128 neurons, a dropout layer to prevent overfitting, and a final dense layer with a sigmoid activation function for binary classification. The total number of parameters is about 3.3 million, all of which are trainable. The training process begins with the first epoch (see output). Next, several warnings appear, such as the *OUT_OF_RANGE* error, which indicates that the dataset or generator ran out of data. This typically happens when the number of steps per epoch exceeds the available batches in the dataset. To resolve this, the dataset should be repeated using the *.repeat()* function. However, despite these warnings, the model continues to train. Throughout the epochs, the training of the model and validation accuracy and loss are displayed. Initially, the model struggles, with erratic accuracy and loss values due to the small dataset and potential overfitting issues. However, as training progresses, the model achieves 100% accuracy on the validation set, which could indicate overfitting rather than generalization. Nevertheless, the number of training images is low so that it can be shown in the pages of this work.

The model is not truly learning to differentiate between the classes but may instead be memorizing the training data. Nevertheless, at the end of the training, another warning appears, suggesting using the native *Keras* format (*model.save('my_model.keras')*) rather than the *HDF5* format (*model.save('my_model.h5')*). However, this is a recommendation for better compatibility with future versions of *Keras* and *TensorFlow*. Once training is over, the model is saved to the specified path (*cat_histology_model.h5*), and a final prediction is made (for testing) on a new image selected from the same histological slide, but a different area unused for training. Thus, the model predicts the image as "tumor" with a confidence of 99.01%. While this confidence level appears high, given the potential overfitting and the limited size of the validation set, the result should be interpreted cautiously. Nonetheless, the console output indicates that the model was successfully trained and saved, but the presence of multiple warnings and the potential for overfitting should prompt the reader to improve the dataset size and variability (i.e., increase the number of training images), adjust the model architecture, or employ better data augmentation techniques (please see the previous chapters on augmentation). *Important observation:* It is clear that the train folder is there to cement in the CNN what it must recognize. But why is there a validation folder? The validation folder needs to contain images because it serves as a separate dataset used to evaluate the performance of the CNN during training. The validation dataset is not used to train the model directly but rather to check how well the model generalizes to unseen data. Thus, by using the validation images, the accuracy of the model, loss, and other metrics can be monitored throughout the training process to prevent overfitting and ensure that the model does not just memorize the training data. This allows for a more reliable assessment of the performance of the model when applied to new, unseen data. Therefore, the validation folder must contain a representative sample of images similar to those in the training set but *not identical*, thus, ensuring the model robustness and generalizability in real-world applications. For the above experiment, only two images were used for the "validation" folder (i.e., that is, in the "normal" and "tumor" subfolders). The reason was to see the importance of the images for the validation folder in the next experiment which was done this time on dog skin (see next subsections).

7.4 Transfer Learning and Pretrained Models

Transfer learning is a powerful machine learning technique that leverages pretrained models to solve new, but related tasks more efficiently and effectively. In the context of microscopy image analysis, transfer learning has gained significant traction due to its ability to achieve high performance even with limited labeled data. Histopathological image analysis often involves complex tasks such as detecting specific cellular morphologies, segmenting tissues, or classifying disease states. These tasks typically require large amounts of annotated data to train deep learning models from scratch, which is both time-consuming and labor-intensive. Transfer learning provides a solution by utilizing models that have been pretrained on large and diverse datasets, such as ImageNet, and then fine-tuning them on domain-specific data. This approach allows researchers to benefit from the knowledge already embedded in these pretrained models, thereby reducing the need for extensive data and computational resources. Pretrained models, such as *VGG*, *ResNet*, *Inception*, and *EfficientNet*, that were mentioned above, have been extensively trained on vast datasets containing millions of images across thousands of categories [164]. These models have learned to recognize and extract hierarchical features from images, such as edges, textures, shapes, and more complex patterns [165]. In transfer learning, the idea is to transfer this knowledge to a new domain, such as histopathological images, where the low- to mid-level features learned from natural images remain highly useful. The application of transfer learning to microscopy image analysis is particularly beneficial due to the high dimensionality and variability of histological images. Tissue samples can exhibit a wide range of staining patterns, morphological variations, and structural complexities. A pretrained model, when fine-tuned on a relatively small but well-annotated dataset, can adapt to these specific features, enabling accurate detection and classification of pathological changes. For example, in the detection of cancerous cells in histopathological slides, a model pretrained on a generic image dataset can be fine-tuned on a dataset of annotated cancer images. The model learns to recognize subtle differences between normal and malignant cells, such as nuclear size, chromatin texture, and tissue organization, thereby improving diagnostic accuracy. In addition to classification tasks, transfer learning is also highly effective for image segmentation, which is critical in quantitative microscopy. Models such as *U-Net*, which are specifically designed for biomedical image segmentation, can be enhanced using transfer learning by initializing the encoder path with weights from a pretrained model. This approach provides a strong foundation for the model to learn the complex spatial relationships between pixels that define cellular structures and tissue boundaries. The use of pretrained encoders can dramatically improve the segmentation performance, particularly in scenarios where labeled data is sparse. Another advantage of transfer learning is its ability to adapt models to different imaging modalities or conditions. For example, models pretrained on histopathological images stained with hematoxylin and eosin (H&E) can be adapted to other staining techniques or even different types of microscopy, such as immunohistochemistry (IHC) or fluorescence microscopy. Transfer learning also supports the concept of domain adaptation, where models trained on data from one institution or scanner can be adapted to data from another institution or scanner with minimal additional training. This adaptability is crucial for creating generalized models that perform well across varied clinical settings, thus reducing the risk of overfitting to specific data distributions and improving the robustness of the model and applicability in real-world scenarios. Moreover, the success of transfer learning in microscopy image analysis is further augmented by the use of techniques such as data augmentation, regularization, and fine-tuning strategies. Data augmentation involves generating new training examples through transformations such as rotation, flipping, scaling, and color adjustments, which help the model generalize better by simulating variability. Regularization methods, such as dropout and weight decay, prevent overfitting during the fine-tuning process, ensuring that the model does not become overly sensitive to the limited training data. Fine-tuning strategies, such as gradually unfreezing layers or using differential learning rates for different layers, provide additional control over the training process, allowing the model to adapt more effectively to new tasks. One of the key considerations when employing transfer learning is the selection of which layers to freeze or train. Lower layers of a pretrained model, which capture general features like edges and textures, are often frozen to preserve their learned representations. Higher layers, which capture more abstract and task-specific features, are typically retrained to adapt to the new domain. The depth of fine-tuning depends on the similarity between the source and target tasks. If the target task is closely related to the original training task, minimal fine-tuning may be required. Conversely, for more dissimilar tasks, deeper fine-tuning may be necessary to adjust the representations of the model to the new data.

7.4.1 An AI-Powered Histological Image Classification Using a Pretrained Model

Above, an implementation was used to train a CNN model to be able to recognize a type of skin tumor in cats. This new implementation provides a user-friendly graphical interface (GUI) to load the pretrained deep learning model and use it for classification of histological images from cat ear tissue as either "normal" or "tumor." Thus, Python Implementation 7.2 serves as an interactive extension of the previous source code, which focused on training a CNN for diagnostic purposes in veterinary histology. Therefore, the purpose of this code is to make it easier for nonprogrammers, such as veterinary pathologists or researchers, to use AI for image analysis without a deep understanding about the underlying technical complexities. Before running this code, the reader must have already trained a model using the previous code and saved it as an *.h5* file, in this case, named *cat_histology_model.h5*.

A pretrained CNN model used for diagnosis

```
import os
import numpy as np
import tkinter as tk
from tkinter import filedialog, Label, Button
from tensorflow.keras.models import load_model
from tensorflow.keras.preprocessing import image
from PIL import Image, ImageTk

# Global variable
# to store the model:
model = None

# Function to load
# the pretrained model:
def load_model_file():
    global model
    # Open a file dialog to select the model
    # file (cat_histology_model.h5 in our case):
    model_path = filedialog.askopenfilename(title="Select Pretrained Model File",
                                            filetypes=[("H5 files", "*.h5")])
    if model_path:
        # Load the model:
        model = load_model(model_path)
        label_status.config(text=f"Model loaded from: {os.path.basename(model_path)}")
        print(f"Model loaded from: {model_path}")
    else:
        print("No model file selected.")

# Function to select
# and classify an image:
def classify_image():
    global model
    if model is None:
        print("Please load a pretrained model first.")
        label_status.config(text="Error: No model loaded!")
        return

    # Open a file dialog to select an image:
    image_path = filedialog.askopenfilename(title="Select an Image for Prediction",
                                            filetypes=[("Image files", "*.jpg *.jpeg *.png")])

    if image_path:
        # Load and preprocess the selected image:
        img = image.load_img(image_path, target_size=(128, 128))
        img_array = image.img_to_array(img)

        # Expand dimensions to match
        # the model input shape:
        img_array = np.expand_dims(img_array, axis=0)
        img_array /= 255.0  # Normalize.

        # Predict using the model:
        prediction = model.predict(img_array)

        # Output the result:
        if prediction[0] > 0.5:
            result_text = f"Tumor: {prediction[0][0] * 100:.2f}% likelihood"
        else:
            result_text = f"Normal: {(1 - prediction[0][0]) * 100:.2f}% likelihood"

        # Display the image and the result:
        img_display = Image.open(image_path)
        img_display = img_display.resize((200, 200))
        img_display = ImageTk.PhotoImage(img_display)
        label_image.config(image=img_display)
        label_image.image = img_display
        label_result.config(text=result_text)
        print(result_text)
    else:
        print("No image file selected.")

# Set up the GUI:
root = tk.Tk()
root.title("AI Diagnostic Tool for Histology Images")

# Load model button:
button_load_model = Button(root, text="Load Pretrained Model", \
    command=load_model_file)
button_load_model.grid(row=0, column=0, padx=10, pady=10)

# Classify image button:
button_classify_image = Button(root, text="Classify Image", \
    command=classify_image)
button_classify_image.grid(row=0, column=1, padx=10, pady=10)

# Status label to show
# which model is loaded:
label_status = Label(root, text="No model loaded.")
label_status.grid(row=1, column=0, columnspan=2, pady=5)

# Label to display
# the selected image:
label_image = Label(root)
label_image.grid(row=2, column=0, columnspan=2, pady=10)

# Label to display the result:
label_result = Label(root, text="Prediction result will appear here.")
label_result.grid(row=3, column=0, columnspan=2, pady=5)

# Run the GUI:
root.mainloop()
```

Python implementation 30: AI-Powered histological image classification using a pretrained model.

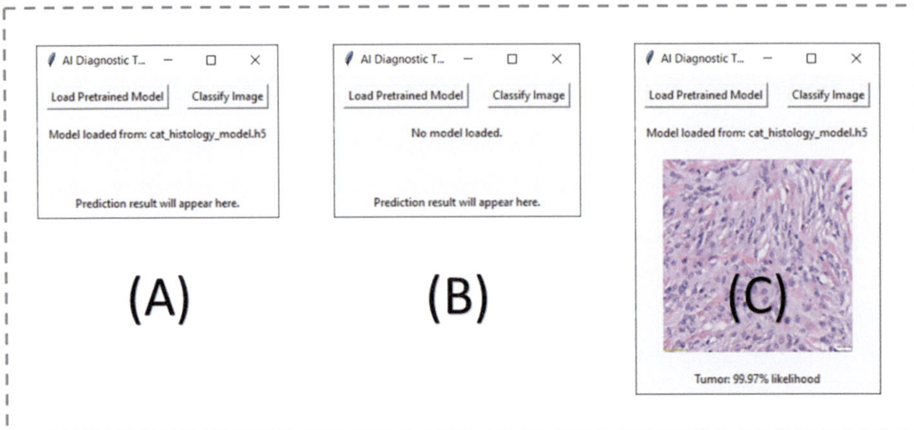

Fig. 7.3 Graphical User Interface (GUI) of the AI Diagnostic Tool for Histology Images. The tool allows users to load a pretrained convolutional neural network (CNN) model and classify histological images as "normal" or "tumor" with a confidence score. (A) The GUI showing a loaded pretrained model (cat_histology_model.h5) ready for image classification. The "Load Pretrained Model" button allows users to select a model file, while the "Classify Image" button enables them to choose an image for analysis. (B) The GUI when no model is loaded; the "Classify Image" button is inactive until a model is loaded. (C) The GUI displaying the prediction result after an image is classified as "tumor" with a 99.97% likelihood, demonstrating the effectiveness of the AI-based diagnostic tool for distinguishing between normal and tumoral cat ear tissue based on histological analysis

This file contains a model trained on 36 histological images from ear tissue of a cat, specifically focused on sarcomas of soft tissue at 40× magnification with a scale bar of 20 μm. The training dataset included 15 images of normal tissue and 21 images of tumoral tissue (one image not shown in Fig. 7.2 for aesthetic reasons) that ensures a decent balanced set for the binary classification task (Figs. 7.1 and 7.2). Note that just like before, the code requires the necessary Python libraries installed, including *TensorFlow*, *Keras*, *numpy*, *PIL* (Pillow), and *tkinter*. The *tkinter* library is used here to create the GUI components, making the code suitable for environments where a graphical interface can be launched (Fig. 7.3).

Python Implementation 7.2 consists of two main functionalities accessible via buttons in the GUI. The first button, "Load Pretrained Model," allows the user to select a model file (such as *cat_histology_model.h5*) from their file system. When the model is loaded successfully, a status message is updated to inform the user of the successful load operation. The loaded model is stored in a global variable model, which is then used for predictions. The second button, "Classify Image," enables the user to select a new histological image from their file system for classification. Once an image is selected, it is preprocessed to match the input requirements of the model (i.e., resized to 128 × 128 pixels and normalized). The loaded model then predicts whether the image shows normal or tumoral tissue. The prediction result is presented both as text with the likelihood percentage and as a visual output where the selected image is displayed on the GUI. The code makes use of the pretrained model file *cat_histology_model.h5*, which encapsulates the learning from the previous training process on histological images. The model was trained to recognize soft tissue sarcomas in cat ear tissue based on features such as cell morphology, tissue organization, and nuclear appearance (Figs. 7.1 and 7.2). This capability is highly valuable for veterinary diagnostic applications, enabling a consistent and potentially more accurate diagnoses compared to manual assessment (i.e., in obvious cases with low difficulty at least). Also, the ratio between the number of images in the training and validation folders is generally recommended to be around 80:20 or 70:30 [163]. This means that 70–80% of the total images should be used for training the model, and the remaining 20–30% should be used for validation [163]. The larger training set allows the model to learn from diverse examples and generalize better, while the validation set is used to monitor the model's performance on unseen data, helping to tune hyperparameters and prevent overfitting. The exact ratio can vary depending on the size of the dataset and the specific problem, but an 80:20 split is a common standard that provides a good balance between training and validation.

7.5 Deep Learning Architectures for Veterinary Pathology

Deep learning architectures have emerged as powerful tools in the analysis of histopathological images, providing automated, high-accuracy methods for detecting and diagnosing diseases in veterinary pathology. Veterinary pathology presents unique challenges due to the multitude of species, tissue types, and disease manifestations. The need for robust, adaptable, and scalable solutions has led to the application of deep learning models, such as *convolutional neural networks* (CNNs), *U-Net*, and more advanced networks like *ResNet*, *DenseNet*, and *vision transformers* (ViTs), specifically made to address these complexities. *Convolutional neural networks* (CNNs) are foundational deep learning models that have revolutionized image analysis by learning hierarchical feature representations directly from input data. In veterinary pathology, CNNs are particularly effective for image classification tasks, such as distinguishing between normal and diseased tissues, identifying specific types of cells, or detecting the presence of parasites in blood smears. The ability of CNNs to learn spatial hierarchies makes them well suited for recognizing patterns and features unique to different species and tissue types. For instance, a CNN can be trained to differentiate between canine, feline, and equine tissue samples, learning the subtle differences in cellular morphology that are critical for accurate diagnosis. CNN architectures such as *AlexNet*, *VGGNet*, and *ResNet* have been applied to veterinary histopathology, achieving high accuracy in classifying various tissue pathologies when trained on large datasets [84, 158]. For more complex tasks that involve localizing and segmenting specific regions within an image, *U-Net* has become a widely adopted architecture. Originally developed for biomedical image segmentation, the encoder-decoder structure of *U-Net* is highly effective in segmenting cellular and subcellular structures in histopathological images. The *U-Net* architecture combines a contracting path to capture context and an expanding path to achieve precise localization, making it ideal for applications where accurate boundaries are crucial, such as in tumor margin detection or identifying regions of necrosis and inflammation. In veterinary pathology, *U-Net* models can be trained to segment different tissue components, such as epithelial cells, connective tissue, and inflammatory infiltrates, which are important for grading tumors or assessing the severity of inflammation. More advanced architectures, such as *ResNet* (residual networks) and *DenseNet* (densely connected convolutional networks), have also been applied to veterinary pathology, particularly in tasks that require deep networks capable of learning intricate patterns from large and complex datasets [166]. *ResNet*, with its residual connections, allows for the training of very deep networks without the vanishing gradient problem, enabling the model to learn more sophisticated features that are essential for distinguishing between closely related tissue types or subtle pathological changes [167]. *DenseNet*, on the other hand, connects each layer to every other layer in a feed-forward fashion, which helps in alleviating the vanishing gradient problem, improves feature propagation and encourages feature reuse. These characteristics make *DenseNet* particularly useful for analyzing images with high variability, such as those encountered in multispecies veterinary datasets where pathological features can differ significantly from one species to another [166, 168]. Another promising deep learning architecture in veterinary pathology is the *vision transformer* (ViT) [169]. Unlike CNNs, which rely on convolutional operations, ViTs use a self-attention mechanism to capture global dependencies within an image. This architecture has shown great promise in natural image classification and is now being adapted for use in medical and veterinary imaging. *ViTs* are especially advantageous when analyzing large histological images that contain a complex arrangement of cellular and tissue structures, as they can model long-range dependencies more effectively than traditional CNNs. For example, *ViTs* can be used to identify patterns of metastasis across entire tissue sections, a task that requires understanding the spatial relationship between distant regions of interest. Therefore, combining deep learning architectures with domain-specific knowledge has also led to the development of hybrid models that integrate multiple networks to leverage their respective strengths. For instance, a hybrid model might combine a CNN for feature extraction with a *U-Net* for segmentation, creating an end-to-end framework that can classify tissue types while simultaneously segmenting regions of interest [170]. Such models are useful in veterinary pathology, where comprehensive analysis often requires both classification (e.g., benign vs. malignant) and segmentation (e.g., identifying tumor margins). Another approach is to integrate deep learning models with classical machine learning techniques, such as *random forests* or *support vector machines* (SVMs), to build ensemble models that are more robust to variability in veterinary datasets [171]. Despite the advancements, the application of deep learning architectures in veterinary pathology is not without challenges. One significant hurdle is the variability in staining, imaging quality, and sample preparation across different veterinary clinics and research institutions, which can affect model performance. Developing robust models that generalize well across such variations requires extensive data augmentation, normalization techniques, and, in some cases, domain adaptation methods [80,

114]. Additionally, the need for interpretability in clinical settings necessitates the development of *explainable AI* (XAI) methods that can provide insights into how models arrive at their decisions [172]. This is critical for gaining the trust of veterinary pathologists and ensuring that the models are used effectively in practice.

7.6 Explainable AI and Model Interpretability

Explainable AI (XAI) and model interpretability have become critical areas of focus in the application of artificial intelligence and machine learning, particularly in sensitive fields such as medicine and histopathology [173]. As machine learning models, especially deep learning models like *convolutional neural networks* (CNNs), become increasingly prevalent in microscopy image analysis, the demand for understanding how these models arrive at their decisions has grown [174]. In clinical settings, trust and transparency are essential. Pathologists and clinicians need to know not only the what of a model's prediction but also the why. This requirement drives the development of techniques that make AI systems more interpretable, thus fostering confidence in their use for diagnostic and prognostic purposes. Traditional machine learning models like *decision trees*, *logistic regression*, and *support vector machines* are relatively easy to interpret because of their simpler structure and fewer parameters. For example, decision trees provide a clear path of reasoning that can be easily visualized and understood. However, deep learning models such as CNNs, which have proven highly effective for image classification, segmentation, and detection tasks, often operate as "black boxes." These models consist of multiple layers with millions of parameters, making it difficult to discern how input features influence the final output. Several methods have been developed to enhance the interpretability of deep learning models in microscopy image analysis. One widely used approach is gradient-weighted *class activation mapping* (Grad-CAM), which provides visual explanations for CNN-based models [157]. Grad-CAM highlights the regions in an input image that are most influential in the decision-making process of the model by computing the gradients of the target output concerning the feature maps of the last convolutional layer [157]. These gradients are then used to generate a heatmap that overlays the original image, indicating the areas that contributed most to the prediction. For instance, in a histopathological image, Grad-CAM can reveal which regions of a tissue sample led the model to classify it as benign or malignant, thereby providing a visual cue that aligns with the interpretive practices of the pathologist [157]. Another approach involves saliency maps, which aim to identify the pixels in an image that have the highest impact on the output of the model. Saliency maps are particularly useful in microscopy image analysis as they can highlight the specific cellular or subcellular structures that are driving the predictions for the model. For example, a saliency map might reveal that the decision of the model to label a region as indicative of a certain type of cancer is driven by the presence of abnormally shaped nuclei or the specific texture of the cytoplasm. This insight can be cross-referenced with established medical knowledge, thus validating the model's decision or revealing areas where the model might be making errors. Another promising technique is *layer-wise relevance propagation* (LRP), which decomposes the prediction of a deep neural network into the contributions of each input feature [175]. Unlike Grad-CAM and saliency maps, which often provide a more general view, LRP can offer a more fine-grained explanation by mapping the relevance of each pixel or small group of pixels to the final prediction. This method is particularly advantageous in distinguishing between subtle differences in tissue samples that may not be obvious to the naked eye but are detected by the model. LRP provides a way to break down the decision-making process of deep learning models, allowing pathologists to understand better how different regions and features within an image contribute to a diagnosis. The integration of these *explainable AI* techniques into microscopy image analysis is essential for several reasons. First, it enhances the transparency of the decision-making process, which is crucial for building trust among clinicians, pathologists, and regulatory bodies. Second, it facilitates collaborative decision-making, where the predictions made by AI models can be used to support, rather than replace, the expertise of human pathologists. For example, in complex cases where a definitive diagnosis is challenging, an explainable model can provide additional insights that help narrow down the possible interpretations or guide further testing. Moreover, *explainable AI* can aid in the iterative refinement of machine learning models [173]. Researchers can identify specific weaknesses or biases in the training process of the model by understanding where and why a model makes incorrect predictions. This feedback loop enables the development of more robust and accurate models, tailored to specific diagnostic challenges or adapted to particular datasets and contexts. It also provides an opportunity to integrate domain knowledge into the model training process, aligning computational predictions with established medical criteria and guidelines. Explainability also has implications for regulatory compliance and ethical considerations in the deployment of AI in medical diagnostics. Regulatory agencies such as the Food and Drug Administration (FDA) in the *United States* and the *European Medicines Agency* (EMA) in the *European Union* are increasingly focusing on the transparency of AI models used in healthcare. Explainable AI techniques provide the necessary documentation and interpretability needed to meet these regulatory standards, ensuring that AI-based diagnostic tools are safe, effective, and trustworthy.

7.7 Case Studies: AI-Powered Diagnostics in Veterinary Medicine

AI-powered diagnostics in veterinary medicine are becoming increasingly important as machine learning and deep learning algorithms advance [176]. These technologies enable the automated analysis of complex histological images and assist in diagnosing various diseases in animals with greater accuracy and speed. The following case studies demonstrate how AI-driven approaches have been successfully applied in veterinary pathology to address some of the most pressing challenges faced by clinicians and researchers. One significant case study involves the use of convolutional neural networks (CNNs) for the early detection of canine mammary tumors [177]. Canine mammary tumors are the most common neoplasms in female dogs and pose a substantial diagnostic challenge due to their varied histopathological presentations [178]. Traditionally, pathologists rely on visual inspection and manual analysis of tissue slides, a process that is time-consuming and prone to variability. In this study, a deep learning model based on CNNs was trained on a large dataset of annotated histopathological images from canine mammary gland tissues [179]. The model was able to distinguish between benign and malignant tumors with high accuracy, demonstrating comparable performance to that of expert pathologists. Moreover, the AI system provided heatmaps that highlighted regions of the image most indicative of malignancy, thereby offering a transparent and interpretable decision-making process. This case study underscores the potential of AI to enhance diagnostic accuracy and consistency, ultimately leading to better prognostic evaluations and treatment planning [180]. Another notable application of AI in veterinary diagnostics is the use of deep learning for the detection and classification of inflammatory and neoplastic diseases in equine tissues [181]. Horses are more susceptible to various inflammatory conditions that can significantly impact their health and performance. Histopathological examination is often required to differentiate between types of inflammation and neoplastic lesions. In this context, a machine learning model was developed to automatically classify histological images of equine tissues into categories such as granulomatous, lymphoplasmacytic, and neutrophilic inflammation, as well as neoplastic conditions like squamous cell carcinoma and lymphoma [181]. The use of a dataset consisting of thousands of annotated images allows the AI model to achieve a high sensitivity and specificity in distinguishing between these conditions. This automated approach provides a powerful tool for equine practitioners, enabling more rapid and accurate diagnosis, which is essential for timely and appropriate treatment interventions. AI-powered diagnostics have also shown promise in the field of veterinary dermatopathology, particularly for diagnosing complex skin conditions in companion animals [182]. Skin diseases in dogs and cats are among the most frequent reasons for veterinary consultations, and they often require histopathological examination to determine the underlying cause [183]. Some studies focused on developing a deep learning model to identify and classify various skin conditions, including autoimmune dermatoses, neoplastic growths, and infectious diseases [184–186]. The model in these cases was trained on high-resolution histological images, and it successfully differentiated between these conditions with remarkable precision. This AI tools not only streamlined the diagnostic workflow but also reduced the likelihood of misdiagnosis, which is crucial given the overlap in histopathological features among different dermatological diseases. The application of AI in the diagnosis of infectious diseases in livestock is another critical area where machine learning is making a substantial impact [187]. For instance, bovine tuberculosis, a significant zoonotic disease, requires accurate and rapid diagnosis to control its spread among cattle herds and prevent transmission to humans [188]. A deep learning approach was employed to analyze histopathological slides of lymph node tissues from cattle, aiming to detect granulomas associated with *Mycobacterium bovis* infection [189]. The AI model was able to identify granulomas with high accuracy, even in early-stage infections where visual detection by pathologists could be challenging. The integration of this AI system into routine diagnostic workflows can facilitate the early detection of bovine tuberculosis, enabling timely interventions and reducing the economic burden on the livestock industry [190]. AI-powered image analysis is also being explored in the context of exotic and zoo animal diagnostics, where the availability of specialized pathologists is limited [191]. For example, a machine learning model was developed to assist in the diagnosis of neoplastic and infectious diseases in reptiles and avian species [192]. These animals often present with unique histopathological features that differ significantly from more common domestic species, making accurate diagnosis challenging. In addition to diagnostic applications, AI is being used to predict treatment responses and outcomes in veterinary oncology [193]. For instance, a study focusing on canine lymphoma utilized machine learning algorithms to analyze histopathological features and molecular markers, predicting the likelihood of treatment response to chemotherapy [194]. All these examples indicate the way veterinary diagnosis and research will look like in the future; thus, more and more advanced machine learning models are expected to fill up these high expectations.

Appendices

Glossary of Terms

Algorithm A step-by-step procedure or formula for solving a problem, typically by a computer.

Artificial Intelligence (AI) The simulation of human intelligence processes by machines, especially computer systems, which includes learning, reasoning, and self-correction.

Brightfield Microscopy Brightfield microscopy is one of the most elementary forms of microscopy illumination techniques available to view a sample. In this technique, light is transmitted directly through or reflected off the sample, causing the specimen to appear dark against a bright background. It is commonly used for stained or naturally pigmented specimens.

Blob Detection A method in image processing intended to detect regions in a digital image that differ in properties, like brightness or color, compared to surrounding regions.

BGR Format BGR format stands for blue, green, red, which is a color representation format commonly used in computer graphics and image processing, particularly in the OpenCV library. In this format, the color channels are arranged in memory in the order of blue first, followed by green, and then red. Each channel typically has a value ranging from 0 to 255, representing the intensity of that color in the image. This format differs from the more commonly used RGB format by the order of channels. BGR is used internally by some libraries because of historical reasons related to the way color information is stored and processed in computer memory.

E. Gagniuc, *Computational Analysis of Microscopy Images*,
https://doi.org/10.1007/978-3-031-86270-0

Class Activation Mapping (CAM) CAM is a visualization technique for identifying important regions in an image that a convolutional neural network (CNN) focuses on for classification. It helps in understanding model decisions, especially in medical imaging.

Confocal Microscopy Confocal microscopy offers increased optical resolution and contrast by using a spatial pinhole to block out-of-focus light in image formation. This technique enables the collection of serial optical sections from thick specimens, providing three-dimensional reconstruction of the sample. It is widely used for the detailed analysis of the structural components of cells within intact tissue sections.

Convolutional Neural Network (CNN) A class of deep neural networks, most commonly applied to analyzing visual imagery, that use a variation of multilayer perceptrons designed to require minimal preprocessing.

Data Augmentation Techniques used to increase the amount of data by adding slightly modified copies of already existing data or newly created synthetic data from existing data.

Deep Learning A subset of machine learning in artificial intelligence that has networks capable of learning unsupervised from data that is unstructured or unlabeled.

Decision Trees Decision Trees are a simple, interpretable algorithm that splits data into branches based on feature values to make predictions. Each branch represents a decision rule, leading to a conclusion at the leaf nodes. They can overfit if not pruned properly.

Electron Microscopy Electron microscopy uses beams of electrons instead of light to create an image of the specimen, achieving much higher resolutions and allowing much greater magnification. This method can provide detailed information about the structure of cells, organelles, and macromolecules. It is crucial in both life sciences and materials science for studying the ultrastructure of a wide array of biological and inorganic specimens.

Feature Extraction The process of defining and calculating the properties of significant information from images, sounds, and other types of data for use in analysis or classification.

Fluorescence Microscopy Fluorescence microscopy uses fluorescence instead of, or in addition to, reflection and absorption to study properties of organic or inorganic substances. A specific wavelength of light is used to excite electrons in molecules of certain compounds and induce them to emit light of a longer wavelength. This technique is extensively used in the field of biological sciences to observe cells and cellular components with high specificity and sensitivity.

Gradient-Weighted Class Activation Mapping (Grad-CAM) A technique for making convolutional neural network decisions transparent by visualizing the specific regions of the input that are important for a given prediction.

Histogram Equalization A method in image processing of contrast adjustment using the histogram of the image.

Image Processing The technique of performing operations on an image to enhance it or extract useful information.

Layer-wise Relevance Propagation (LRP) LRP is an interpretability technique that assigns relevance scores to each input feature by backpropagating the prediction through the network, showing how each feature contributed to the decision. Useful for transparent AI in critical fields like healthcare.

Logistic Regression It is a binary classification algorithm that models the probability of a class using a logistic function. It outputs probabilities that can be used for classification and is effective for problems where classes are linearly separable.

Machine Learning The scientific study of algorithms and statistical models that computer systems use to perform tasks by relying on patterns and inference.

Morphological Operations Nonlinear operations related to the shape or morphology of features in an image, such as dilation, erosion, opening, and closing.

Noise Reduction The process of removing noise from a signal. In imaging, this includes various filters and techniques to smooth the appearance of the image.

Phase-Contrast Microscopy Phase-contrast microscopy enhances the contrast of transparent and colorless objects by influencing the optical path of light. This method is particularly useful for examining living cells and organisms, allowing for the detailed observation of cellular structures that would otherwise be invisible under standard bright-field illumination.

Quantitative Analysis The examination of measurable and verifiable data such as lengths, widths, quantities, and densities.

Random Forests Random Forests is an ensemble learning method that builds multiple decision trees and combines their outputs for more accurate and robust classification or regression. It reduces overfitting by averaging results and is effective for high-dimensional data.

Region-Based Segmentation A technique in image processing that partitions an image into regions which are similar according to a set of predefined criteria.

ResNet A type of convolutional neural network (CNN) model that uses skip connection or shortcut to jump over some layers, which helps in solving the vanishing gradient problem in deep networks.

RGB Format RGB format stands for red, green, blue, which is a color representation model widely used in computer graphics, digital imaging, and displays. In the RGB format, each pixel is represented by a combination of red, green, and blue color channels, each having a value ranging from 0 to 255. The combination of these three channels creates a broad spectrum of colors. The RGB format is the standard format for most imaging applications, including digital cameras, computer screens, and Web graphics, because it directly corresponds to the way colors are perceived by the human eye. In the RGB format, the channels are stored in memory in the order of red, followed by green, and then blue.

Staining The technique of adding a dye to a specimen to enhance the visibility of tissues, cells, or cellular components under a microscope.

Support Vector Machines (SVM) A supervised machine learning model that uses classification algorithms for two-group classification problems. It is effective in high-dimensional spaces and can handle nonlinear data using kernel functions.

Transfer Learning The reuse of a pretrained model on a new problem. It is currently very popular in the field of deep learning because it can train deep neural networks with comparatively little data.

U-Net A type of convolutional neural network that is used primarily for image segmentation. Its architecture is named for its U-shaped symmetry.

Watershed Algorithm An image processing technique used for image segmentation. It is used to find and mark the watershed lines in digital images.

Z-stack In microscopy, a collection of images taken at different focal depths to create a 3D representation of the sample.

Different Types of Microscopic Images

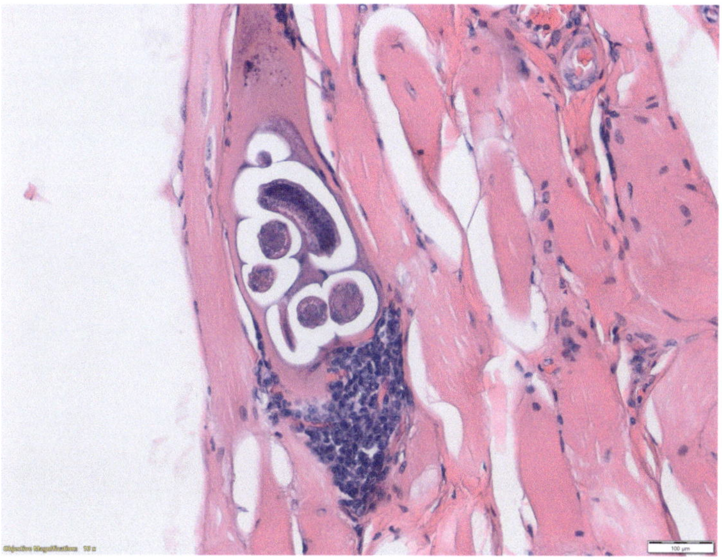

Histological image showing *Trichinella* spp. larvae encysted in the striated muscle tissue of a dog. The section is stained with hematoxylin and eosin (H&E), which highlights the different tissue components: nuclei appear dark blue or purple, and cytoplasm and muscle fibers are pink. The encysted larvae of *Trichinella* spp. are clearly visible within the muscle fibers, surrounded by inflammatory infiltrates composed predominantly of lymphocytes, plasma cells, macrophages, and eosinophils, indicating a parasitic infestation response. The surrounding muscle fibers show signs of degeneration and reactive changes due to the presence of the parasite. The objective magnification is 10×, with a scale bar representing 100 μm, providing context for the size and distribution of the pathological findings within the muscular tissue. This image illustrates the characteristic morphological features of *Trichinella* spp. infestation in muscle tissue, which is important for veterinary diagnosis and understanding the pathological processes in parasitic life cycle

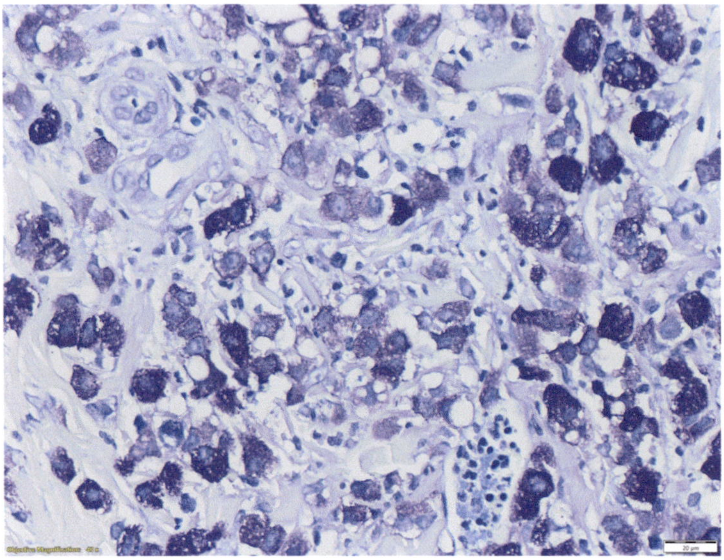

Histological image of a canine cutaneous mast cell tumor (mastocytoma) from a palpebral location, stained with toluidine blue, a special stain used to highlight mast cells. The image is captured at 40× magnification, with a scale bar indicating 20 µm. Mast cells are stained in dark blue to purple hues, showing metachromatic granules, which are characteristic of mast cell tumors. The surrounding stroma and connective tissue appear in lighter blue shades, providing contrast that allows for clear visualization of the mast cells. The high magnification allows for detailed observation of the cellular morphology, including nuclear features and cytoplasmic granularity, which are important for diagnostic confirmation and assessment of the degree of malignancy. This staining technique is critical in veterinary pathology for identifying mast cells due to their distinctive staining properties, aiding in the diagnosis of cutaneous mastocytomas in canine patients

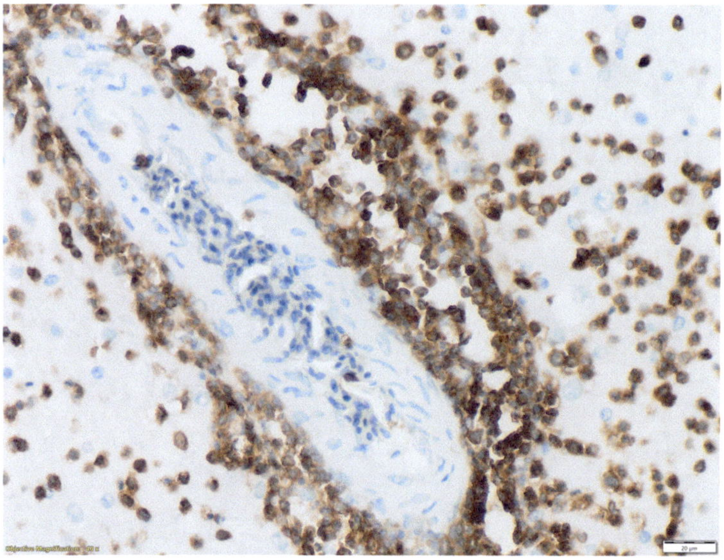

Histological image of a chicken brain tissue sample, demonstrating immunohistochemical staining using anti-CD3 antibodies to detect T lymphocytes. The tissue is from a case of Marek's disease, a viral-induced lymphoma common in poultry. The image was captured at 40× magnification, with a scale bar indicating 20 μm. The brown staining indicates the presence of CD3-positive T cells, which are densely packed around a blood vessel, a hallmark of Marek's disease–associated lymphoid infiltration. The blue counterstaining highlights the nuclei of other cells within the tissue, providing contrast to better visualize the distribution and morphology of T lymphocytes. This type of immunohistochemical staining is critical in the diagnosis and study of Marek's disease, as it allows for the identification and localization of lymphocyte cell populations within affected tissues

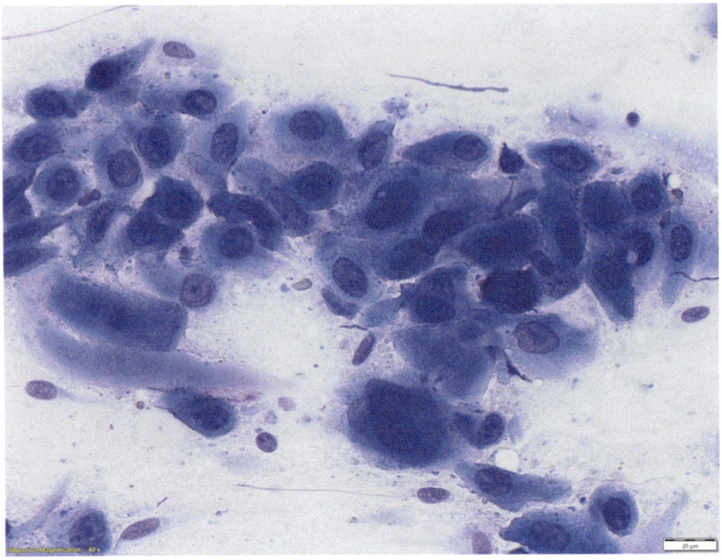

Cytological image of squamous cell carcinoma, obtained from a skin sample taken from the ear of a cat, stained with May-Grünwald Giemsa and captured at 40× magnification. The slide is stained to highlight cellular and nuclear details, with a scale bar representing 20 μm. The image reveals clusters of neoplastic squamous epithelial cells with prominent nuclei and nucleoli. Anisokaryosis and anisocytosis are characteristic features of squamous cell carcinoma. The cells exhibit varying sizes and shapes, with evidence of nuclear pleomorphism and hyperchromasia, which are indicative of malignancy

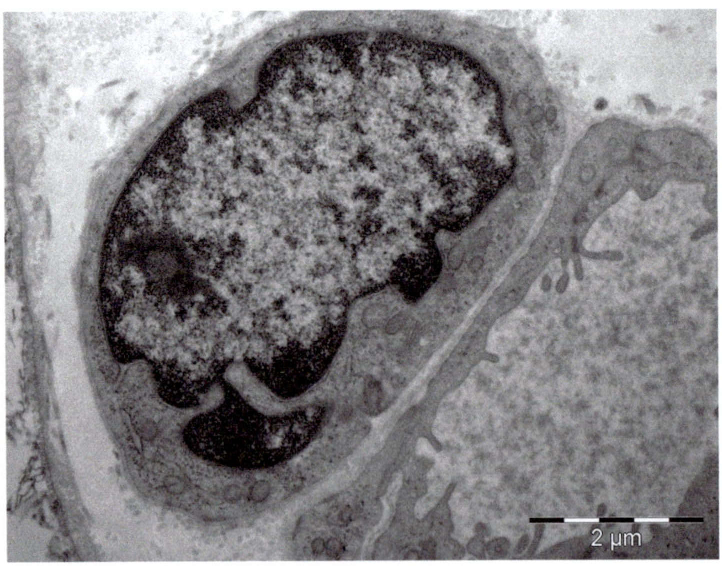

Electron microscopy image showing a white adipocyte from a cow, with a scale bar of 2 μm. The image highlights the ultrastructure of the adipocyte nucleus, characterized by marginal heterochromatin and

central euchromatin. The surrounding cytoplasm contains organelles such as mitochondria and the nucleus, which appear compressed against the cell membrane due to the large size of the lipid droplet

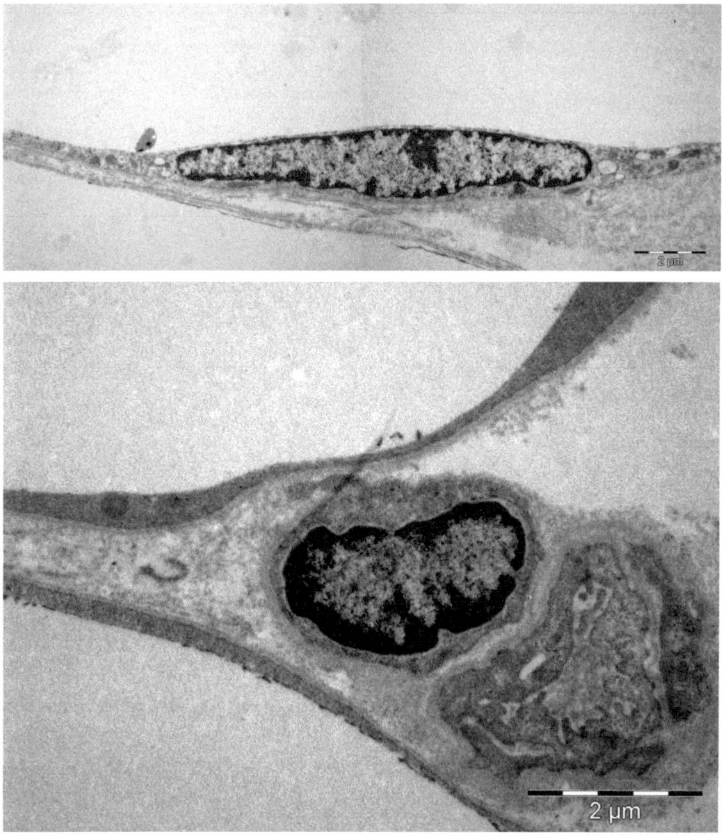

Electron microscopy image of a white adipocyte from a cow, with a scale bar of 2 µm. The image displays the ultrastructural features typical of a white adipocyte, including a large central lipid droplet that dominates the cell's interior, surrounded by a thin layer of cytoplasm. Visible organelles, such as mitochondria, are scattered within the cytoplasmic rim, providing insight into the metabolic activities within these fat cells

Confocal microscopy image of a moth head, captured using a Nikon AX R confocal microscope with NSPARC technology. The image reveals intricate details of the head anatomy of the moth (*Tineola bisselliella*), including its compound eyes, sensory organs, and mouthparts. The use of confocal imaging allows for a sharp and high-contrast visualization of different structures, providing clear differentiation between various tissue types and cellular components

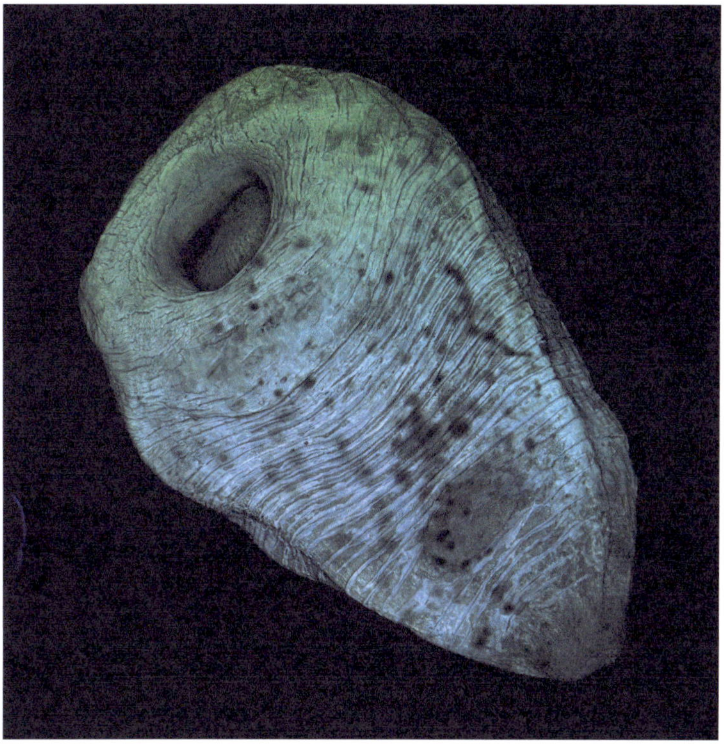

Confocal microscopy image of *Paramhistomum* spp. captured using a Nikon AX R confocal microscope with NSPARC technology. The image provides a detailed view of the structural features of *Paramhistomum*, a parasitic trematode known to infect the digestive tracts of ruminants. The confocal imaging technique allows for precise visualization of both the external morphology (i.e., the tegument)

Acknowledgments

I would like to thank Dr. Paul A. Gagniuc for the technical support and the *ELTA 90 Medical Reaserch* company through the representative of Mr. Adrian Dragne for the access to the Nikon AX R confocal microscope with NSPARC technology (the last pictures in the appendices). I also want to thank Prof. Dr. Gherghiceanu Mihaela, for the access to the electron microscope at the *National Institute of Pathology Victor Babeş*.

References

1. K. Thompson, P. Rodriguez, *Specialized Veterinary Techniques: From Neurology to Cardiology* (Springer, New York, 2019)
2. D. Schaffer, J. Foster, Veterinary medicine and public health. J. Vet. Sci. Med. Diag **7**(2), 1–12 (2018)
3. K. Nguyen, A. Lee, Food safety and the critical role of veterinary medicine. Food Control **69**(1), 112–119 (2016)
4. L. Martin, H. Brown, The integration of digital imaging and biochemical analyses in veterinary practice. Vet. Q. **40**(1), 55–67 (2020)
5. E. Harper, M. Clarkson, *Ethics and Professional Practice in Veterinary Medicine* (Elsevier, London, 2018)
6. T. Sanders, *Holistic Veterinary Medicine: New Approaches to Health and Healing* (Cambridge University Press, Cambridge, 2021)
7. J. Pawley, *Handbook of Biological Confocal Microscopy* (Springer, New York, 2006)
8. P. Thévenaz, M. Unser, Optimization of mutual information for multiresolution image registration. IEEE Trans. Image Process. **9**(12), 2083–2099 (2007)
9. R.C. Gonzalez, R.E. Woods, *Digital Image Processing*, 3rd edn. (Prentice Hall, Upper Saddle River, 2008)
10. G. Litjens, T. Kooi, B.E. Bejnordi, A.A.A. Setio, F. Ciompi, M. Ghafoorian, J.A.W.M. van der Laak, B. van Ginneken, C.I. Sánchez, A survey on deep learning in medical image analysis. Med. Image Anal. **42**(1), 60–88 (2017)
11. A. Webb, *Statistical Pattern Recognition*, 2nd edn. (Wiley, Hoboken, 2005)
12. D.L. Pham, C. Xu, J.L. Prince, Current methods in medical image segmentation. Annu. Rev. Biomed. Eng. **2**(1), 315–337 (2000)
13. I.N. Bankman, *Handbook of Medical Image Processing and Analysis*, 2nd edn. (Academic Press, San Diego, 2008)
14. J.A. Kiernan, *Histological and Histochemical Methods: Theory and Practice* (Scion, Bloxham, 2015)
15. J. Goldstein, D.E. Newbury, D.C. Joy, C.E. Lyman, P. Echlin, E. Lifshin, L. Sawyer, J.R. Michael, *Scanning Electron Microscopy and X-ray Microanalysis* (Springer, New York, 2003)
16. J.D. Bancroft, M. Gamble, *Theory and Practice of Histological Techniques* (Churchill Livingstone, Edinburgh, 2008)
17. J.B. Pawley, *Handbook of Biological Confocal Microscopy* (Springer, Boston, 2006)

E. Gagniuc, *Computational Analysis of Microscopy Images*,
https://doi.org/10.1007/978-3-031-86270-0

18. K. Murphy, M. Smuts, A. Kotze, Advances in veterinary pathology through digital and computational methods. Vet. Pathol. **55**(4), 472–481 (2018)
19. J.C. Russ, R.T. Dehoff, *Practical Stereology* (Springer, New York, 2020)
20. M.J. Day, The Immunopathology of Animal Diseases: The Role of Microscopy. Vet. Pathol. **51**(2), 32–43 (2014)
21. B. Jones, A. Vale, *The Comprehensive History of Veterinary Medicine* (Oxford University Press, Oxford, 2017)
22. E.G. Ruestow, *The Microscope in the Dutch Republic: The Shaping of Discovery* (Cambridge University Press, Cambridge, 1996)
23. P. Harting, *History of the Microscope* (John Van Voorst, London, 1855)
24. K.R. Porter, M.A. Bonneville, *An Introduction to the Fine Structure of Cells and Tissues* (Lea & Febiger, Philadelphia, 1963)
25. B.J. Ford, *Single Lens: The Story of the Simple Microscope* (Harper & Row, New York, 1985)
26. B. Alberts, A. Johnson, J. Lewis, et al., *Molecular Biology of the Cell*, 4th edn. (Garland Science, New York, 2002)
27. A.M. Silverstein, *Paul Ehrlich's Receptor Immunology: The Magnificent Obsession* (Academic Press, San Diego, 2002)
28. G.L. Turner, *Nineteenth Century Scientific Instruments* (University of California Press, Berkeley, 1989)
29. L.G. Shapiro, G.C. Stockman, *Computer Vision* (Prentice Hall, Upper Saddle River, 2001)
30. D. McMullan, Electron microscopy: 50 years of development in imaging and analysis. Adv. Electron. Electron Phys. **75**(1), 271–358 (1988)
31. J.D. Webster, R.W. Dunstan, Whole slide imaging for educational purposes. J. Vet. Med. Educ. **41**(1), 27–36 (2014)
32. R.D. Goldman, D.L. Spector, *Live Cell Imaging: A Laboratory Manual* (Cold Spring Harbor Laboratory Press, Cold Spring Harbor, 2005)
33. R.H. Webb, Confocal optical microscopy. Annu. Rev. Biomed. Eng. **8**(1), 407–422 (1996)
34. Y. LeCun, Y. Bengio, G. Hinton, Deep learning. Nature **521**(7553), 436–444 (2015)
35. G. Litjens, T. Kooi, B.E. Bejnordi, et al., A survey on deep learning in medical image analysis. Med. Image Anal. **42**(1), 60–88 (2017)
36. S. Inoué, K.R. Spring, *Video Microscopy: The Fundamentals*, 2nd edn. (Plenum Press, New York, 1997)
37. D.B. Murphy, M.W. Davidson, *Fundamentals of Light Microscopy and Electronic Imaging* (Wiley-Blackwell, Hoboken, 2012)
38. B. Alberts, A. Johnson, J. Lewis, M. Raff, K. Roberts, P. Walter, *Molecular Biology of the Cell*, 6th edn. (Garland Science, New York, 2014)
39. T. Wilson, Resolution and optical sectioning in the confocal microscope. J. Microsc. **244**(2), 113–121 (2011)
40. M. Abramowitz, M.W. Davidson, *Handbook of Optical and Laser Scanning* (CRC Press, Boca Raton, 2017)
41. B. Huang, M. Bates, X. Zhuang, Super-resolution fluorescence microscopy. Annu. Rev. Biochem. **78**(1), 993–1016 (2009)
42. D.J. Stephens, V.J. Allan, Light microscopy techniques for live cell imaging. Science **300**(5616), 82–86 (2003)
43. J. Jonkman, C.M. Brown, G.D. Wright, K.I. Anderson, A.J. North, Tutorial: guidance for quantitative confocal microscopy. Nat. Protoc. **15**(5), 1585–1611 (2020)
44. C.H. Fox, F.B. Johnson, J. Whiting, P.P. Roller, Formaldehyde fixation. J. Histochem. Cytochem. **33**(8), 845–853 (1985)

45. D. Hopwood, Theoretical and practical aspects of glutaraldehyde fixation. Histochem. J. **1**(3), 267–276 (1969)
46. J.A. Kiernan, Formaldehyde, formalin, paraformaldehyde and glutaraldehyde: what they are and what they do. Microsc. Today **8**(1), 8–12 (2000)
47. R.D. Lillie, H.M. Fullmer, Histopathologic technic and practical histochemistry. Arch. Pathol. Lab Med. **100**(12), 658 (1976)
48. J.D. Bancroft, M. Gamble, *Theory and Practice of Histological Techniques*, 6th edn. (Churchill Livingstone, New York, 2008)
49. R.W. Horobin, J.A. Kiernan, Conn's biological stains: a handbook of dyes, stains and fluorochromes for use in biology and medicine. Biotech. Histochem. **77**(5–6), 237–275 (2002)
50. J.J. Bozzola, L.D. Russell, *Electron Microscopy: Principles and Techniques for Biologists* (Jones & Bartlett Learning, Boston, 1999)
51. J.K.C. Chan, *Tissue-Based Diagnostics: Pathology and Histology* (Springer, London, 2014)
52. D.L. Spector, R.D. Goldman, *Cells: A Laboratory Manual* (Cold Spring Harbor Laboratory Press, Cold Spring Harbor, 2006)
53. A.D. Weston, L. Hood, Systems biology, proteomics, and the future of health care: toward predictive, preventative, and personalized medicine. J. Proteome Res. **3**(2), 179–196 (2004)
54. S.A. Rice, R.D. McCall, *Digital Imaging: A Practical Approach* (American University Press, New York, 2012)
55. P.J. Sudharshan, C. Petitjean, F. Spanhol, L.E. Oliveira, L. Heutte, P. Honeine, Machine learning for medical image analysis: a survey. Pattern Recogn. **93**(1), 25–47 (2019)
56. B. Pritt, J. Harmon, Microscopy in veterinary medicine: applications in diagnostics and research. Vet. Pathol. **49**(3), 376–389 (2012)
57. M. Maxie, *Jubb, Kennedy & Palmer's Pathology of Domestic Animals*, vol 1 (Elsevier, St. Louis, MO, 2015)
58. D. Haines, E. Clark, Immunohistochemical detection of Leptospira serovars in formalin-fixed, paraffin-embedded tissues. J. Vet. Diagn. Invest. **3**(4), 376–379 (1991)
59. J. Zachary, *Pathologic Basis of Veterinary Disease* (Elsevier, St. Louis, 2017)
60. M. McGavin, J. Zachary, *Pathologic Basis of Veterinary Disease Expert Consult* (Elsevier, St. Louis, 2013)
61. C. Culling, R. Allison, W. Barr, *Cellular Pathology Technique* (Butterworth-Heinemann, London, 1985)
62. J. Cullen, N. MacLachlan, J. Collins, *Veterinary Histology and Microscopy: A Diagnostic Approach* (CABI, Wallingford, 2020)
63. B.S. Pritt, J.F. Harmon, Artifacts in histopathology. Vet. Pathol. **49**(3), 608–618 (2012)
64. M.L. Watson, Staining of tissue sections for electron microscopy with heavy metals. J. Biophys. Biochem. Cytol. **4**(4), 475–478 (1958)
65. W.E. Grizzle, J.L. Fredenburgh, Avoiding artifacts in the preparation of surgical pathology specimens. Hum. Pathol. **32**(8), 857–860 (2001)
66. L. Milios, L. Sompayrac, Observer variability in the diagnosis of histopathology specimens. J. Clin. Pathol. **68**(3), 157–164 (2015)
67. A.C. Ritts, S. Orlov, Computational microscopy and its applications to veterinary medicine. Vet. Pathol. **50**(6), 1060–1070 (2013)
68. U. Giger, L. Cornetta, Veterinary education and training in the age of technological advances. J. Vet. Med. Educ. **45**(4), 635–644 (2018)
69. J.H. Smith, D. Dabbs, Digital pathology: current status and future perspectives. Vet. Pathol. **53**(1), 6–17 (2016)
70. G. Dougherty, *Digital Image Processing for Medical Applications* (Cambridge University Press, Cambridge, 2013)

71. M. Tolnay, T. Dzieciatkowski, Advances in veterinary digital imaging techniques. Vet. Clin. N. Am. Small Anim. Pract. **45**(2), 281–293 (2015)
72. P.E. Ginn, L.L. Donaldson, Histopathology in veterinary dermatology: a guide for the practitioner. Vet. Clin. N. Am. Small Anim. Pract. **40**(1), 11–35 (2010)
73. A. Esteva, B. Kuprel, R.A. Novoa, J. Ko, S.M. Swetter, H.M. Blau, S. Thrun, Dermatologist-level classification of skin cancer with deep neural networks. Nature **542**(7639), 115–118 (2017)
74. G. Campanella, M.G. Hanna, L. Geneslaw, A. Miraflor, V. Werneck Krauss Silva, K.J. Busam, T.J. Fuchs, Clinical-grade computational pathology using weakly supervised deep learning on whole slide images. Nat. Med. **25**(8), 1301–1301 (2019)
75. K. Sirinukunwattana, S.E.A. Raza, Y.W. Tsang, D.R.J. Snead, I.A. Cree, N.M. Rajpoot, Locality sensitive deep learning for detection and classification of nuclei in routine colon cancer histology images. IEEE Trans. Med. Imaging **35**(5), 1196–1206 (2016)
76. W. Denk, H. Horstmann, Serial block-face scanning electron microscopy to reconstruct three-dimensional tissue nanostructure. PLoS Biol. **2**(11), e329 (2004)
77. M. Kiupel, Tumor pathology of domestic animals. Vet. Pathol. **54**(2), 309–311 (2017)
78. G. Campanella, M.G. Hanna, L. Geneslaw, A. Miraflor, V.W.K. Silva, K.J. Busam, E. Brogi, V.E. Reuter, D.S. Klimstra, T.J. Fuchs, Clinical-grade computational pathology using weakly supervised deep learning on whole slide images. Nat. Med. **25**(8), 1301–1309 (2019)
79. K. Bera, K.A. Schalper, D.L. Rimm, V. Velcheti, A. Madabhushi, Artificial intelligence in digital pathology—new tools for diagnosis and precision oncology. Nat. Rev. Clin. Oncol. **16**(11), 703–715 (2019)
80. Y. Xu, T. Mo, Q. Feng, P. Zhong, M. Lai, E.I. Chang, Deep learning of feature representation with multiple instance learning for medical image analysis. IEEE Trans. Neural Netw. Learn. Syst. **29**(11), 2624–2635 (2016)
81. L. Pantanowitz, G.M. Quiroga-Garza, L. Bien, R. Heled, D. Laifenfeld, C. Linhart, T.W. Bauer, An artificial intelligence algorithm for prostate cancer diagnosis in whole slide images of core needle biopsies: a blinded clinical validation and deployment study. Lancet Digit. Health **2**(8), e407–e416 (2020)
82. J. van der Laak, G. Litjens, F. Ciompi, Deep learning in histopathology: the path to the clinic. Nat. Med. **27**(5), 775–784 (2021)
83. D. Komura, S. Ishikawa, Machine learning methods for histopathological image analysis. Comput. Struct. Biotechnol. J. **16**(1), 34–42 (2018)
84. M. Aubreville, C.A. Bertram, T.A. Donovan, A. Maier, Machine learning in veterinary pathology: utility and guidelines. Vet. Pathol. **58**(5), 822–842 (2021)
85. K. Suzuki, Overview of deep learning in medical imaging. Radiol. Phys. Technol. **10**(3), 257–273 (2017)
86. M.N. Gurcan, L.E. Boucheron, A.M. Can, A. Madabhushi, N. Rajpoot, B. Yener, histopathological image analysis: a review. IEEE Rev. Biomed. Eng. **2**(1), 147–171 (2009)
87. J.M.S. Prewitt, M.L. Mendelsohn, The analysis of cell images. Ann. N. Y. Acad. Sci. **128**(3), 1035–1053 (1966)
88. F.A. Spanhol, L.S. Oliveira, C. Petitjean, L. Heutte, A dataset for breast cancer histopathological image classification. IEEE Trans. Biomed. Eng. **63**(7), 1455–1462 (2016)
89. N. Otsu, A threshold selection method from gray-level histograms. IEEE Trans. Syst. Man Cybern. **9**(1), 62–66 (1979)
90. T. Fawcett, An introduction to ROC analysis. Pattern Recogn. Lett. **27**(8), 861–874 (2006)
91. C.A. Schneider, W.S. Rasband, K.W. Eliceiri, NIH image to ImageJ: 25 years of image analysis. Nature Methods **9**(7), 671–675 (2012)
92. A.E. Carpenter, T.R. Jones, M.R. Lamprecht, C. Clarke, I.H. Kang, O. Friman, D.A. Guertin, J.H. Chang, R.A. Lindquist, J. Moffat, P. Golland, D.M. Sabatini, CellProfiler: image analysis software for identifying and quantifying cell phenotypes. Genome Biol. **7**(10), R100 (2006)

93. P. Bankhead, M.B. Loughrey, J.A. Fernández, Y. Dombrowski, D.G. McArt, P.D. Dunne, S. McQuaid, R.T. Gray, L.J. Murray, H.G. Coleman, J.A. James, M. Salto-Tellez, P.W. Hamilton, QuPath: open source software for digital pathology image analysis. Sci. Rep. **7**(1), 16878 (2017)
94. H. Attia, H. Ibrahim, Review on applications of MATLAB in image processing. Indones. J. Electr. Eng. Comput. Sci. **5**(1), 1–11 (2017)
95. P. Gagniuc, *Coding Examples from Simple to Complex – Applications in MATLAB* (Springer, Cham, 2024), pp. 1–255
96. A. Paszke, S. Gross, F. Massa, A. Lerer, J. Bradbury, G. Chanan, T. Killeen, Z. Lin, N. Gimelshein, L. Antiga, A. Desmaison, A. Köpf, E. Yang, Z. DeVito, M. Raison, A. Tejani, S. Chilamkurthy, B. Steiner, L. Fang, S. Chintala, PyTorch: an imperative style, high-performance deep learning library. Adv. Neural Inf. Proces. Syst. **32**(1), 8024–8035 (2019)
97. M. Abadi, P. Barham, J. Chen, Z. Chen, A. Davis, J. Dean, M. Devin, S. Ghemawat, G. Irving, M. Isard, M. Kudlur, J. Levenberg, R. Monga, S. Moore, D. G. Murray, B. Steiner, P. Tucker, V. Vasudevan, P. Warden, X. Zheng, TensorFlow: a system for large-scale machine learning, in *12th USENIX Symposium on Operating Systems Design and Implementation (OSDI 16), Savannah, GA*, 2016, pp. 265–283
98. J.S. Krueger, D.L. Rimm, Digital pathology and machine learning—recent advances in histopathology and molecular diagnosis. Cell. Mol. Med. Rep. **1**(2), 49–55 (2017)
99. H. Wäckerle, ZEN 2012: Microscopy software from Carl Zeiss. Nat. Methods **9**(7), 677 (2012)
100. S. Martínez-Subiela, A. Tvarijonaviciute, J. Cerón, Use of biomarkers in veterinary medicine: from theory to practice. Vet. J. **247**(1), 58–65 (2019)
101. S. Smith, C. Brown, Comparative anatomy in veterinary medicine: implications for disease diagnosis and treatment. Vet. Pathol. **56**(5), 709–717 (2019)
102. J. Koenig, C. Wenzel, S. Forster, et al., Multispecies applications of artificial intelligence in veterinary diagnostics. Comput. Biol. Med. **142**(1), 105189 (2022)
103. A.B. Smith, C.D. Johnson, Comparative computational pathology: species-specific approaches to liver biopsy analysis. J. Vet. Inf. **35**(2), 123–135 (2020)
104. R.S. Thompson, M.L. Baker, Species differences in lymphoma: a comparative study of canine and feline cases. Vet. Oncol. **44**(1), 89–102 (2019)
105. B. Broeckx, E. Stock, F. Farnir, et al., Artificial intelligence in veterinary medicine: where are we now? J. Vet. Med. Educ. **46**(3), 316–328 (2019)
106. M. Denwood, S. Holden, M. Brennan, et al., The challenges of standardizing veterinary imaging data for computational analysis. J. Vet. Diagn. Invest. **32**(6), 890–899 (2020)
107. J. Vandeweerd, N. Kirschvink, P. Clegg, S. Vandenput, Is evidence-based medicine adequate for the challenges of veterinary practice? Vet. J. **193**(2), 197–202 (2012)
108. P. Perona, J. Malik, Scale-space and edge detection using anisotropic diffusion. IEEE Trans. Pattern Anal. Mach. Intell. **12**(7), 629–639 (1990)
109. E. Gagniuc, M. Siteavu, E. Pitoiu, E. Ciobotaru-Pîrvu, Encephalitis associated with PCV3 infection in a Romanian pig farm. J. Comp. Pathol. **203**(1), 61–61 (2023)
110. M. Siteavu, E. Gagniuc, L. Darie, E. Pitoiu, E. Ciobotaru-Pîrvu, M. Taulescu, First report of porcine circovirus type-3 in Romanian pig farms: pathological and molecular investigations. J. Comp. Pathol. **191**(1), 49–49 (2022)
111. J.S. Lee, Digital image smoothing and the sigma filter. Comput. Vis. Graphics Image Process. **24**(2), 255–269 (1983)
112. A. Buades, B. Coll, J. Morel, A non-local algorithm for image denoising, in *Proceedings of the IEEE Computer Society Conference on Computer Vision and Pattern Recognition (CVPR), San Diego, CA, USA*, vol. 2, (IEEE, Piscataway, 2005), pp. 60–65
113. G. Shorten, T. Khoshgoftaar, B. Furht, A survey on image data augmentation for deep learning. J. Big Data **6**(1), 60–60 (2019)

114. G. Shorten, T.M. Khoshgoftaar, B. Furht, A survey on image data augmentation for deep learning. J. Big Data **6**(1), 60 (2019)
115. P. Gagniuc, C. Ionescu-Tîrgoviște, C.H. Rădulescu, Automatic growth detection of cell cultures through outlier techniques. Int. J. Comput. Commun. **8**(3), 407–415 (2013)
116. P. Gagniuc, *Coding Examples from Simple to Complex – Applications in Python* (Springer, Cham, 2024), pp. 1–245
117. L. Vincent, P. Soille, Watersheds in digital spaces: an efficient algorithm based on immersion simulations. IEEE Trans. Pattern Anal. Mach. Intell. **13**(6), 583–598 (1991)
118. C. Ionescu-Tirgoviste, P. Gagniuc, E. Gubceac, L. Mardare, I. Popescu, S. Dima, M. Militaru, A 3D map of the islet routes throughout the healthy human pancreas. Sci. Rep. **5**(1), 14634 (2015)
119. C. Glasbey, An analysis of histogram-based thresholding algorithms. Graphical Models Image Process. **55**(6), 532–537 (1993)
120. J. Canny, A computational approach to edge detection. IEEE Trans. Pattern Anal. Mach. Intell. **PAMI-8**(6), 679–698 (1986)
121. R.M. Haralick, S.R. Sternberg, X. Zhuang, Image analysis using mathematical morphology. IEEE Trans. Pattern Anal. Mach. Intell. **PAMI-9**(4), 532–550 (1987)
122. R. Gonzalez, R. Woods, Sobel operators and their use in edge detection, in *Digital Image Processing*, 4th edn., (Prentice Hall, Upper Saddle River, 2018), pp. 213–214
123. B. De, S. Saha, S. Saha, Application of Sobel edge detection for diagnosis of brain tumors using image processing techniques, in *IEEE Symposium on Computational Intelligence in Healthcare and e-health (CICARE), Orlando, FL*, (IEEE, Piscataway, 2014), pp. 1–6
124. T.F. Chan, L. Vese, Active contours without edges. IEEE Trans. Image Process. **10**(2), 266–277 (2001)
125. R.M. Haralick, K. Shanmugam, I. Dinstein, Textural features for image classification. IEEE Trans. Syst. Man Cybern. **SMC-3**(6), 610–621 (1973)
126. T. Ojala, M. Pietikäinen, T. Mäenpää, Multiresolution gray-scale and rotation invariant texture classification with local binary patterns. IEEE Trans. Pattern Anal. Mach. Intell. **7**(971–987), 24 (2002)
127. A. Jain, F. Farrokhnia, Unsupervised texture segmentation using Gabor filters. Pattern Recogn. **24**(12), 1167–1186 (1991)
128. H. Tang, E. Gonzalez, L. Kleinman, M. Wall, Shape-based feature extraction for histopathological image classification. Comput. Biol. Med. **99**(1), 1–12 (2018)
129. N. Petkov, M. Viergever, Adaptive filtering for enhanced texture discrimination. IEEE Trans. Pattern Anal. Mach. Intell. **20**(5), 533–539 (1998)
130. H. Greenspan, B. van Ginneken, R. Summers, Guest editorial deep learning in medical imaging: overview and future promise of an exciting new technique. IEEE Trans. Med. Imaging **35**(5), 1153–1159 (2016)
131. E.C. Jones, B.I. Reiner, Automated quantitative image analysis in pathology: the future is now. Diagn. Pathol. **14**(1), 113–113 (2019)
132. J. Pallud, L. Capelle, L. Taillandier, D. Fontaine, E. Mandonnet, R. Guillevin, H. Duffau, Quantitative imaging with apparent diffusion coefficient (ADC) and contrast enhancement predicts survival in patients with diffuse low-grade gliomas. Neuro-Oncology **12**(11), 1088–1099 (2010)
133. S. Robinson, S. Patel, R. Davis, C. Hough, Quantitative analysis of histological staining: a review of automated techniques. J. Histochem. Cytochem. **64**(3), 131–145 (2016)
134. A. Zare, T. Kangarloo, P. Mansoori, M. Makki, A. Salimi, M. Amini, Automated measurement of cellular morphological parameters in histological images using image analysis techniques. J. Microsc. **270**(1), 56–65 (2018)

135. E. Gubceac (Condrut), in *Research on the Morphology and Pathology of Adipose Tissue in Some Animal Species*, ed. by M. Manuella, (University of Agricultural Sciences and Veterinary Medicine Bucharest, Faculty of Veterinary Medicine, Department of Anatomical Pathology, Bucharest, 2014), p. 232
136. H.M. Kaltenbach, S. Emmert-Buck, C. Prosser, Quantitative analysis of nuclear morphology in histopathology. Vet. Pathol. **58**(2), 289–301 (2021)
137. C.M. Olsen, D. Mowery, E. Clement, Automated morphological analysis of tissue images: methods and applications. J. Pathol. Inf. **9**(1), 45–52 (2018)
138. G. Maicas, C. Zheng, G. Litjens, N. Rajpoot, I. Cree, A. Madabhushi, A. Lee, Morphological parameterization and classification in histopathological images using machine learning. IEEE Trans. Med. Imaging **37**(5), 1288–1300 (2018)
139. P. Gagniuc, C. Ionescu-Tîrgovişte, Gene promoters show chromosome specificity and reveal chromosome territories in humans. BMC Genomics **14**(1), 278 (2013)
140. P. Gagniuc, C. Ionescu-Tirgoviste, Eukaryotic genomes may exhibit up to 10 generic classes of gene promoters. BMC Genomics **13**(1), 512 (2012)
141. P. Gagniuc, P. Cristea, R. Tuduce, C. Ionescu-Tîrgovişte, L. Gavrila, DNA patterns and evolutionary signatures obtained through Kappa Index of Coincidence. Rev. Roum. Sci. Techn. Électrotechn. et Énerg. **57**(1), 100–109 (2012)
142. C. Ionescu-Tirgoviste, M.G.E. Ioan Valentin, M. Militaru, D. Gutu, D. Lixandru, A cytomorphometric analysis of adipocytes from the omental and abdominal subcutaneous adipose tissue. Proc. Rom. Acad. Ser. B Chem. Life Sci. Geosci. **13**(3), 212–236 (2011)
143. E. Condrut, L. Gaita, D. Parvan, M. Militaru, Fractal characteristics of adipocytes in mice fed with a hypercaloric diet. J. Comp. Pathol. **152**(1), 71 (2015)
144. J. Cohen, Statistical power analysis for the behavioral sciences. Appl. Probab. Stat. **39**(3), 467–471 (1988)
145. S. Winkler, A. Basu, S. Datta, H. Dutta, Applying Bayesian methods in veterinary medicine: a review. Stat. Med. **33**(9), 1363–1379 (2014)
146. M. Pepe, The statistical evaluation of medical tests for classification and prediction. Stat. Med. **31**(3), 217–225 (2012)
147. T. Hastie, R. Tibshirani, J. Friedman, The elements of statistical learning: data mining, inference, and prediction. Annu. Rev. Biomed. Eng. **13**(1), 345–373 (2011)
148. K. Baggerly, K. Coombes, D. Barry, Quantitative data analysis in medical imaging. IEEE Trans. Med. Imaging **27**(2), 283–294 (2008)
149. T. Anderson, *An Introduction to Multivariate Statistical Analysis*, Wiley Series in Probability and Statistics, 3rd edn. (Wiley-Interscience, New York, 2003)
150. I. Jolliffe, J. Cadima, Principal component analysis: a review and recent developments. Philos. Trans. R. Soc. A Math. Phys. Eng. Sci. **374**(2065), 20150202 (2016)
151. H. Hotelling, Analysis of a complex of statistical variables into principal components. J. Educ. Psychol. **24**(6), 417–441 (1933)
152. C. Chu, Y. Luo, Y. Zhang, X. Huang, Y. Xia, Application of U-Net CNN for brain tumor detection and segmentation. IEEE Access **8**(1), 58818–58828 (2020)
153. I. Goodfellow, J. Pouget-Abadie, M. Mirza, B. Xu, D. Warde-Farley, S. Ozair, A. Courville, Y. Bengio, Generative adversarial networks. Adv. Neural Inf. Process. Syst **27**(1), 2672–2680 (2014)
154. C. Ionescu-Tîrgovişte, R. Vasilescu, R. Timar, C. Guja, P. Gagniuc, E. Băcanu, R. Gligor, S. Carniciu, The proinsulin-to-adiponectin ratio could be the best practical indicator of the early (prehyperglycaemic) phase of type 2 diabetes. Adipobiology **4**(1), 41–50 (2012)
155. C. Guja, L. Guja, P.A. Gagniuc, C. Ionescu-Tîrgovite, Landscape of monogenic diabetes in the third millenium. Proc. Rom. Acad., Ser. B **15**(3), 217–232 (2013)

156. D. Lixandru, E. Băcanu, B. Vîrgolici, H. Vîrgolici, A. Petruţa, M. Băcanu, P. Gagniuc, C. Ionescu-Tîrgovişte, C. Serafinceanu, Changes in the serum proinflammatory cytokines in patients with elevated Homa-Ir and Type 2 diabetes mellitus. Farmacia **63**(1), 132–139 (2015)
157. R. Selvaraju, M. Cogswell, A. Das, R. Vedantam, D. Parikh, D. Batra, Grad-CAM: visual explanations from deep networks via gradient-based localization, in *Proceedings of the IEEE International Conference on Computer Vision, ICCV*, no. 1, (IEEE, Piscataway, 2017), pp. 618–626
158. E. Shelhamer, J. Long, T. Darrell, Fully convolutional networks for semantic segmentation. IEEE Trans. Pattern Anal. Mach. Intell. **39**(4), 640–651 (2017)
159. P.K. Diederik, B. Jimmy, Adam: a method for stochastic optimization, in *3rd International Conference for Learning Representations*, San Diego, 2015
160. I. Goodfellow, Y. Bengio, A. Courville, Chapter 5: Linear models for classification, in *Deep Learning*, (MIT Press, Cambridge, 2016), pp. 180–184
161. C.M. Bishop, Chapter 1: Introduction (Section 1.3 on model assessment and selection), in *Pattern Recognition and Machine Learning*, (Springer, New York, 2006), pp. 32–36
162. A. Churbanov, A. Wilcox, T. Yuen, E. Georganas, R. Smeliansky, Accelerating deep learning workloads on Intel® architecture using oneAPI Deep Neural Network Library (oneDNN), in *Proceedings of the 2020 IEEE International Parallel and Distributed Processing Symposium Workshops (IPDPSW), New Orleans, Louisiana, USA*, (IEEE, Piscataway, 2020)
163. F. Chollet, Chapter 5: Deep learning for computer vision, in *Deep Learning with Python*, (Manning Publications, Shelter Island, 2018), pp. 153–155
164. C. Szegedy, V. Vanhoucke, S. Ioffe, J. Shlens, Z. Wojna, Rethinking the inception architecture for computer vision, in *Proceedings of the IEEE Conference on Computer Vision and Pattern Recognition, CVPR*, no. 1, (IEEE, Piscataway, 2016), pp. 2818–2826
165. A. Krizhevsky, I. Sutskever, G.E. Hinton, ImageNet classification with deep convolutional neural networks. Commun. ACM **60**(6), 84–90 (2017)
166. Z. Liu, Z. Lin, W. Cao, L. Zhang, Y. Xu, J. Yang, DenseNet-based deep learning models for the automated diagnosis of canine dermatoses from histopathological images. Vet. Dermatol. **31**(5), 415–423 (2020)
167. Z. Wang, X. Luo, K. Chen, H. Zhang, H. Sun, X. Zhao, Application of residual networks (ResNet) in veterinary medicine for the diagnosis of canine lymphoma. Vet. Pathol. **58**(4), 567–576 (2021)
168. Y. Wang, Y. Zhang, X. Shi, Z. Tang, Using DenseNet architecture for classification of canine osteosarcoma in radiographic images. J. Vet. Diagn. Invest. **33**(2), 231–239 (2021)
169. Z. Wu, Y. Li, X. Zhang, Vision transformers in veterinary medicine: a comparative study of CNNs and ViTs for animal disease detection. Vet. Res. Commun. **46**(1), 123–134 (2022)
170. O. Ronneberger, P. Fischer, T. Brox, U-Net: convolutional networks for biomedical image segmentation, in *Medical Image Computing and Computer-Assisted Intervention—MICCAI 2015*, no. 1, vol. 9351, (Springer, Cham, 2015), pp. 234–241
171. C. Cortes, V. Vapnik, Support-vector networks. Mach. Learn. **20**(3), 273–297 (1995)
172. S. Finlayson, I. Chung, A. Kohane, D. Beam, Adversarial attacks against medical deep learning systems. Science **363**(6433), 1287–1289 (2019)
173. A. Holzinger, C. Biemann, C. Pattichis, D. Kell, What do we need to build explainable AI systems for the medical domain? Nat. Rev. AI **1**(9), 520–530 (2017)
174. A. Adadi, M. Berrada, Peeking inside the black-box: a survey on explainable artificial intelligence (XAI). IEEE Access **6**(1), 52138–52160 (2018)
175. S. Bach, A. Binder, G. Montavon, F. Klauschen, K. Müller, W. Samek, On pixel-wise explanations for non-linear classifier decisions by layer-wise relevance propagation. PLoS One **10**(7), e0130140 (2015)

176. G. Campanella, M. Hanna, L. Geneslaw, A. Miraflor, V. Werneck Krauss Silva, K. Busam, T. Fuchs, Clinical-grade computational pathology using weakly supervised deep learning on whole slide images. Nat. Med. **25**(8), 1301–1309 (2019)
177. S. Kim, J. Kim, H. Chung, J. Park, Deep learning-based histopathological classification of canine mammary tumors. Vet. Pathol. **56**(4), 589–597 (2019)
178. C. VandeWalle, K. Franklin, K. Blum, S. Sun, R. Ranganath, Application of convolutional neural networks in the diagnosis of canine mammary gland tumors. J. Vet. Diagn. Invest. **32**(7), 925–932 (2020)
179. N. Pedersen, H. Liu, H. Park, A. Fung, Automated histopathological analysis of canine mammary tumors using deep learning. Vet. Comp. Oncol. **18**(1), 64–73 (2021)
180. A. Riddle, T. Allen, R. Goldstein, Artificial intelligence in veterinary dermatopathology: applications and challenges. Vet. Dermatol. **31**(6), 459–467 (2020)
181. N. Jeurissen, W. De Vos, E. van der Vekens, Machine learning models for classifying inflammatory and neoplastic conditions in equine tissues. Vet. Pathol. **58**(4), 515–526 (2021)
182. D.F. Clark, L. Gomez, A. Williams, R.C. Flynn, Artificial intelligence in veterinary dermatopathology: applications for complex skin conditions in companion animals. J. Vet. Dermatol. **32**(3), 215–223 (2020)
183. A. Khodadoust, S. Kavousi, R. Shalbaf, F. Moradi, Deep learning-based histopathological classification of equine skin lesions. Vet. Pathol. **58**(2), 247–255 (2021)
184. D. Klein, M. Adams, C. Bishop, Machine learning in equine veterinary medicine: a review of current applications and future directions. J. Vet. Sci. **22**(4), 567–576 (2020)
185. F. González, L. Lara, A. Rodrigues, J. Braga, N. Filgueira, C. Figueiredo, Deep learning-based classification of dermatological conditions in canine and feline skin lesions. Vet. Dermatol. **31**(5), 335–346 (2020)
186. S. Patel, R. Agarwal, S. Pillai, V. Menon, Artificial intelligence in veterinary dermatopathology: a pilot study for the diagnosis of skin diseases. J. Vet. Diagn. Invest. **32**(3), 492–501 (2020)
187. M. McCulloch, R. Mitchell, A. Forshaw, R. Barr, D. Fitzpatrick, D. Tobin, Automated detection of granulomas in histopathological slides for the diagnosis of bovine tuberculosis using deep learning. Comput. Biol. Med. **124**(1), 103956 (2020)
188. A. Alonso, F. Pérez-Alonso, J. Benitez, Histopathological analysis for bovine tuberculosis diagnosis using artificial intelligence. Vet. Res. Commun. **44**(3), 209–220 (2020)
189. S. Abdollahi, M. Ghaznavi, A. Zare-Mirakabadi, M. Sepehri, Application of machine learning to diagnose mycobacterium Bovis infection from lymph node slides in cattle. Vet. Immunol. Immunopathol. **212**(1), 35–41 (2019)
190. R. Karim, R. Shakur, M. Rahman, AI-driven approaches for early detection of zoonotic tuberculosis in livestock: a review. Anim. Biotechnol. **31**(2), 121–133 (2021)
191. R. Mandal, M.S. Uddin, A. Ghosh, M. Debnath, AI-assisted diagnostics in exotic animal medicine: a case study of reptiles and birds. Vet. Res. Commun. **45**(3), 369–383 (2021)
192. S. Ashraf, A. Zaki, A. Memon, Machine learning approaches for diagnosing infectious and neoplastic diseases in reptiles and avian species. J. Zoo Wildl. Med. **50**(4), 867–875 (2019)
193. L. Brown, P. Wang, L. Black, J. Trent, Predictive modeling in veterinary oncology: machine learning for prognosis and treatment outcome in canine lymphoma. J. Vet. Intern. Med. **34**(5), 1900–1912 (2020)
194. S. Choi, S. Lee, J. Kim, Artificial intelligence in veterinary oncology: predicting chemotherapy response in canine lymphoma using histopathological and molecular data. Vet. Oncol. **10**(2), 112–123 (2022)
195. R. Selvaraju, A. Das, R. Vedantam, M. Cogswell, D. Parikh, D. Batra, Grad-CAM: Visual Explanations from Deep Networks via Gradient-Based Localization, in *Proceedings of the IEEE International Conference on Computer Vision, ICCV*, no. 1, (IEEE, Piscataway, 2017), pp. 618–626

Index

E. Gagniuc, *Computational Analysis of Microscopy Images*,
https://doi.org/10.1007/978-3-031-86270-0

MIX
Papier aus verantwortungsvollen Quellen
Paper from responsible sources
FSC® C105338

If you have any concerns about our products,
you can contact us on
ProductSafety@springernature.com

In case Publisher is established outside the EU,
the EU authorized representative is:
Springer Nature Customer Service Center GmbH
Europaplatz 3, 69115 Heidelberg, Germany

Printed by Libri Plureos GmbH
in Hamburg, Germany